21世纪本科院校土木建筑类创新型应用人才培养规划教材

土木工程专业英语

主　编　宿晓萍　赵庆明
副主编　沙　勇　李　莹
参　编　吴春利　常　虹

北京大学出版社
PEKING UNIVERSITY PRESS

内 容 简 介

本书共 26 课，每课包括正文和参考译文两大部分，每课编配生词表和常用词组与专业短语，并配有练习题、必要的翻译注释，以及专业基础知识的相关扩展内容和阅读材料，便于学生更好地掌握本书的教学内容，在学习专业知识的同时，可积累大量的相关专业英语词汇，了解科技文章的语法特点与翻译技巧。

本书选材范围涉及建筑工程、道路与桥梁工程、岩土工程、地下工程、建筑设计与构造、工程管理等方面的内容，与土木工程专业课程的基本教学内容结合紧密。全书以文字为主，适当配图，选取的英文文章专业性强，内容浅显易懂，专业词汇丰富。

本书适合作为建筑工程、道路桥梁工程、岩土工程、管理工程等专业的中外合作办学项目、"卓越计划"及应用型本科的教材。

图书在版编目(CIP)数据

土木工程专业英语/宿晓萍，赵庆明主编 . —北京：北京大学出版社，2017.6
(21 世纪本科院校土木建筑类创新型应用人才培养规划教材)
ISBN 978-7-301-28317-2

Ⅰ. ①土… Ⅱ. ①宿… ②赵… Ⅲ. ①土木工程—英语—高等学校—教材 Ⅳ. ①TU

中国版本图书馆 CIP 数据核字(2017)第 112992 号

书　　　名	土木工程专业英语 TUMU GONGCHENG ZHUANYE YINGYU
著作责任者	宿晓萍　赵庆明　主编
责 任 编 辑	伍大维
标 准 书 号	ISBN 978-7-301-28317-2
出 版 发 行	北京大学出版社
地　　　址	北京市海淀区成府路 205 号　100871
网　　　址	http://www.pup.cn　新浪微博：@北京大学出版社
电 子 信 箱	pup_6@163.com
电　　　话	邮购部 010-62752015　发行部 010-62750672　编辑部 010-62750667
印 刷 者	北京圣夫亚美印刷有限公司
经 销 者	新华书店
	787 毫米×1092 毫米　16 开本　18.25 印张　423 千字 2017 年 6 月第 1 版　2021 年 7 月第 3 次印刷
定　　　价	40.00 元

未经许可，不得以任何方式复制或抄袭本书之部分或全部内容。
版权所有，侵权必究
举报电话：010-62752024　电子信箱： fd@pup.pku.edu.cn
图书如有印装质量问题，请与出版部联系，电话：010-62756370

前　　言

中国目前正在进行大规模的基础设施建设，一批大型、复杂、高端的工程项目吸引了国外建筑公司或事务所参与设计、施工与管理等，同时，越来越多的中国建筑企业也涌向国际建筑市场，积极拓展国际工程承包业务。随着国际合作工程项目的日益增多，这对中国土木工程专业人才的质量及其国际合作能力与竞争力提出了新的考验。同时，随着我国"卓越工程师培养计划"的不断推进，高等工程教育已逐渐趋于国际化，高等院校也在不断开展国际合作办学，通过借鉴和利用国外高等教育的优质资源与成功经验，培养具有国际化高素质的专业人才，以增强我国人才在国际工程市场上的竞争能力。

本书从培养应用型国际化土木工程专业人才的目标出发，结合学生出国进修学习及毕业后的工作实际，通过一定数量的土木工程类英语原文文章，训练学生阅读与翻译本专业英文资料的初步能力，掌握一定量的专业英语词汇，为学生出国进修学习夯实专业英语应用基础，提高学生未来工作岗位所需要的专业英语知识和技能。

本书由宿晓萍、赵庆明担任主编，由沙勇、李莹担任副主编，吴春利、常虹参编。本书共分26课，具体编写分工如下：长春工程学院宿晓萍编写第1、2、3、4、5课，长春工程学院赵庆明编写第7、9、15、16、20课，长春工程学院沙勇编写第10、11、12、13、14、17、18、19课，长春工程学院李莹编写第21、22、23、24、25、26课，吉林大学吴春利编写第6课，吉林建筑大学常虹编写第8课。

本书的编写得到了长春工程学院外国语学院高岩松老师的大力支持与帮助，在此表示衷心的感谢。

由于编者水平有限，书中难免有不妥之处，恳请广大读者批评指正。

编　者
2016 年 12 月

目 录

Lesson 1	Civil Engineering	1
Lesson 2	Basic Knowledge of Drawing	11
Lesson 3	Building and Building System	31
Lesson 4	Materials in Construction	44
Lesson 5	Concrete	58
Lesson 6	Introduction to Mechanics of Materials	70
Lesson 7	Site Surveys	78
Lesson 8	Soil Mechanics	89
Lesson 9	Shallow Foundations Forms	98
Lesson 10	Structural Concepts	112
Lesson 11	The Forces on a Building and the Effects	122
Lesson 12	Structural Design	133
Lesson 13	Masonry Structure	143
Lesson 14	Reinforced Concrete Structure	149
Lesson 15	Steel Members	155
Lesson 16	Steel Connections	169
Lesson 17	Tall Buildings	180
Lesson 18	Prestressed Concrete	188
Lesson 19	What Happens to Structure When the Ground Moves	197

Lesson 20   **Underground Space Utilization** ········· 206

Lesson 21   **How Tunnels Are Built** ················· 218

Lesson 22   **Types of Bridges Ⅰ** ···················· 230

Lesson 23   **Types of Bridges Ⅱ** ···················· 241

Lesson 24   **Hydrology and Hydraulics** ············· 252

Lesson 25   **Road Engineering** ······················ 262

Lesson 26   **Traffic Engineering** ···················· 272

参考文献 ··· 281

Lesson 1
Civil Engineering

Civil engineering, the oldest of the engineering specialties, is the planning, design, construction, and management of the built environment.① This environment includes all structures built according to scientific principles, from irrigation and drainage systems to rocket-launching facilities.

Civil engineers build roads, bridges, tunnels, dams, harbors, power plants, water and sewage systems, hospitals, schools, mass transit, and other public facilities essential to modern society and large population concentrations. They also build privately owned facilities such as airports, railroads, pipelines, skyscrapers, and other large structures designed for industrial, commercial, or residential use. In addition, civil engineers plan, design, and build complete cities and towns, and more recently have been planning and designing space platforms to house self-contained communities.

The word "civil" derives from the Latin for citizen. In 1782, an Englishman John Smeaton used the term to differentiate his nonmilitary engineering work from that of the military engineers who predominated at the time②. Since then, the term civil engineering has often been used to refer to engineers who build public facilities, although the field is much broader.

1. Scope

Because it is so broad, civil engineering is subdivided into a number of technical specialties③. Depending on the type of project, the skills of many kinds of civil engineer specialists may be needed.

When a project begins, the site is surveyed and mapped by civil engineers who locate utility placement water, sewer, and power lines. Geotechnical specialists perform soil experiments to determine if the earth can bear the weight of the project. Environmental specialists study the project's impact on the local area: the potential for air and groundwater pollution, the project's impact on local animal and plant life, and how the project can be designed to meet government requirements aimed at protecting the environment. Transportation specialists determine what kind of facilities is needed to ease the burden on local roads and other transportation networks that will result from the completed project.④ Meanwhile, structural specialists use preliminary data to make detailed designs, plans, and specifications for the project. Supervising and coordinating the work of these civil engineer specialists, from the beginning to the end of the project, are the construction management specialists. Based on information supplied by the other specialists, construction management civil engineers estimate quantities and costs of materials and labor, schedule all work, order materials and equipment for the job, hire contractors and subcontractors, and perform other supervisory work to ensure the project is completed on time and as specified.

Throughout any given project, civil engineers make extensive use of computers. Computers are used to design the project's various elements (computer-aided design, or CAD) and to manage it. Computers are a necessity for the modern civil engineer because they permit the engineer to efficiently handle the large quantities of data needed in determining the best way to construct a project.

2. Structural engineering

In this specialty, civil engineers plan and design structures of all types, including bridges, dams, power plants, supports for equipment, special structures for offshore projects, the United States space program, transmission towers, giant astronomical and radio telescopes, and many other kinds of projects. Using computers, structural engineers determine the forces a structure must resist: its own weight, wind and hurricane forces, temperature changes that expand or contract construction materials, and earthquakes. They also determine the combination of appropriate materials: steel, concrete, plastic, stone, asphalt, brick, aluminum, or other construction materials.

3. Water resources engineering

Civil engineers in this specialty deal with all aspects of the physical control of water. Their projects help prevent floods, supply water for cities and for irrigation, manage and control rivers and water runoff, and maintain beaches and other waterfront facilities. In addition, they design and maintain harbors, canals, and locks, build huge hydroelectric dams and smaller dams and water impoundments of all kinds, help design offshore structures, and determine the location of structures affecting navigation.

4. Geotechnical engineering

Civil engineers who specialize in this field analyze the properties of soils and rocks that support structures and affect structural behavior. They evaluate and work to minimize the potential settlement of buildings and other structures that stems from the pressure of their weight on the earth. These engineers also evaluate and determine how to strengthen the stability of slopes and fills and how to protect structures against earthquakes and the effects of groundwater.

5. Environmental engineering

In this branch of engineering, civil engineers design, build, and supervise systems to provide safe drinking water and to prevent and control pollution of water supplies, both on the surface and underground. They also design, build, and supervise projects to control or eliminate pollution of the land and air. These engineers build water and wastewater treatment plants, and design air scrubbers and other devices to minimize or eliminate air pollution caused by industrial processes, incineration, or other smoke-producing activities. They also work to control toxic and hazardous wastes through the construction of special dump sites or the neutralizing of toxic and hazardous substances. In addition, the engineers design and manage sanitary landfills to prevent pollution of surrounding land.

6. Transportation engineering

Civil engineers working in this specialty build facilities to ensure safe and efficient movements of both people and goods. They specialize in designing and maintaining all types of transportation facilities, highways and streets, mass transit systems, railroads and airfields, ports and harbors. Transportation engineers apply technological knowledge as well as consideration of the economic, political, and social factors in designing each project. They work closely with urban planners, since the quality of the community is directly related to the quality of the transportation system.

7. Pipeline engineering

In this branch of civil engineering, engineers build pipelines and related facilities which transport liquids, gases, or solids ranging from coal slurries (mixed coal and water) and semi-liquid wastes, to water, oil, and various types of highly combustible and noncombustible gases. The engineers determine pipeline design, the economic and environmental impact of a project on regions it must traverse, the type of materials to be used — steel, concrete, plastic, or combinations of various materials — installation techniques, methods for testing pipeline strength, and controls for maintaining proper pressure and rate of flow of materials being transported. When hazardous materials are being carried, safety is a major consideration as well.

8. Construction engineering

Civil engineers in this field oversee the construction of a project from beginning to end. Sometimes called project engineers, they apply both technical and managerial skills, including knowledge of construction methods, planning, organizing, financing, and operating construction projects. They coordinate the activities of virtually everyone engaged in the work: the surveyors, workers who lay out and construct the temporary roads and ramps, excavate for the foundation, build the forms and pour the concrete, and workers who build the steel framework. These engineers also make regular progress reports to the owners of the structure.

9. Community and urban planning

Those engaged in this area of civil engineering may plan and develop communities within a city, or entire cities. Such planning involves far more than engineering consideration; environmental, social, and economic factors in the use and development of land and natural resources are also key elements. These civil engineers coordinate planning of public works along with private development. They evaluate the kinds of facilities needed, including streets and highways, public transportation systems, airports, port facilities, water-supply and wastewater-disposal systems, public buildings, parks, and recreational and other facilities to ensure social and economic as well as environmental well-being.

10. Photogrametry, surveying, and mapping

The civil engineers in this specialty precisely measure the Earth's surface to obtain reliable information for locating and designing engineering projects. This practice often involves high-technology methods such as satellite and aerial surveying, and computer processing of photographic imagery. Radio signals from satellites, scans by laser and sonic beams, are converted to maps to provide far more accurate measurements for boring tunnels, building highways and dams, plotting flood control and irrigation projects, locating subsurface geologic formations that may affect a construction project, and a host of other building uses.

11. Other specialties

<u>Two additional civil engineering specialties that are not entirely within the scope of civil engineering but are essential to the discipline are engineering management and engineering teaching.</u>⑤

1) Engineering management

Many civil engineers choose careers that eventually lead to management. Others are able to start their careers in management positions. The civil engineer-manager combines technical knowledge with an ability to organize and coordinate worker power, materials, machinery, and money. These engineers may work in government—municipal, county, state, or federal; in the U. S. Army Corps of Engineers as military or civilian management engineers; or in semiautonomous regional or city authorities or similar organizations. They may also manage private engineering firms ranging in size from a few employees to hundreds.

2) Engineering teaching

The civil engineer who chooses a teaching career usually teaches both graduate and undergraduate students in technical specialties. Many teaching civil engineers engage in basic research that eventually leads to technical innovations in construction materials and methods. Many also serve as consultants on engineering projects, or on technical boards and commissions associated with major projects.

Ⅰ. New Words

1. irrigation *n.* 灌溉
2. drainage *n.* 排水，排水系统，污水
3. sewage *n.* 污水，下水道，污物
4. predominate *vt.* 居支配地位，统治，（数量上）占优势
5. subdivide *vt.* 把……再分，把……细分
6. geotechnical *adj.* 岩土工程技术的

7. specification *n.* 规格，说明书，技术要求
8. supervise *vt.* 监督，管理，指导
9. coordinate *vt.* 协调，调整，整合
10. subcontract *vt.* 转包，分包；*n.* 转包合同，分包合同
11. subcontractor *n.* 转包商，分包者
12. supervisory *adj.* 监督的，管理的
13. asphalt *n.* 沥青，柏油；*vt.* 铺沥青于……
14. aluminum *n.* ［化］铝
15. impoundment *n.* 蓄水，贮水量，围住，扣留
16. scrubber *n.* 洗涤器，滤清器，刷子，擦洗者
17. incineration *n.* 焚化，烧尽，火葬
18. toxic *adj.* 有毒的，中毒的
19. slurry *n.* 泥浆，水泥浆，煤泥
20. combustible *adj.* 易燃的，燃烧性的；*n.* 可燃物，易燃物
21. ramp *n.* 斜坡，坡道，斜面
22. excavate *vt.* 挖掘，开凿
23. aerial *adj.* 空气的，大气的，航空的
24. sonic *adj.* 声音的，音速的，声波的
25. plot *vt.* 测绘，标图，标航路；*n.* 测绘，标图
26. municipal *adj.* 市政的，市立的，地方自治的

Ⅱ. Phrases and Expressions

1. civil engineering   土木工程
2. structural engineering   结构工程
3. water resources engineering   水利资源工程
4. geotechnical engineering   岩土工程
5. environmental engineering   环境工程
6. transportation engineering   交通（运输）工程
7. pipeline engineering   管道工程
8. construction engineering   建筑工程，施工工程
9. engineering management   工程管理
10. drainage system   排水系统

Ⅲ. Notes

① "the oldest of the engineering specialties" 在句中的成分是 civil engineering 的主语补足语。另外，科技文章常用一般现在时表述，注重的是客观事实与真理。

② "differentiate...from" 译为 "与……相区别，不同于……"。句中的 that 指代 work，以避免用词重复。

③ 主句采用被动语态。被动语态的使用是专业英语中常见的语法特点，因为科技人员往往更关心事实和行为，而不是行为者；而且被动语态比主动语态简短，将最重要的信息放在句首，一下就可抓住读者的注意力。

④ "what kind of … project" 是 determine 的宾语从句。在此宾语从句中，"that will result from the completed project" 是 burden 的定语从句。复杂长句多是科技文章的一个显著特点，如主从句、一主多从、并列句应用较多，不仅包含的信息量大，也是为了达到准确描述事物的目的。

⑤ "… not … but …" 意为"不是……而是……"，带有选择的意思。

Ⅳ. Exercises

Fill in the blanks with the information given in the text.

1. Two additional civil engineering specialties that are not entirely within the scope of civil engineering but are essential to the discipline are _____ and _____ .

2. Translate the following expressions into English.
(1) 土木工程_____ (2) 结构工程_____ (3) 岩土工程_____
(4) 管理工程_____ (5) 建筑设计_____ (6) 民用建筑_____

Ⅴ. Expanding

Remember the following terms related to the type of engineering.

1. residential building　居住建筑
2. public building　公共建筑
3. heating and ventilating and air-conditioning engineering/HVAC engineering　采暖与通风工程
4. water and sewerage engineering　给排水工程
5. municipal project　市政工程
6. subway engineering　地铁工程

Ⅵ. Reading Material

Becoming a Civil Engineer

In the English-speaking countries, unlike Continental Europe, a professional engineer who wishes to be fully qualified, must join at least one engineering institution. All these institutions require candidates for admission to prove that they have some years of useful practical experience as an engineer. Each institution is a learned society not unlike a club except that the candidate's strict examination for membership is based mainly on his engineering knowledge, and all institutions publish engineering literature in their own subjects, usually in their own subjects, usually in their monthly journal. Each has several grades of membership, from the highest, full Member, down through the usual grade,

Associate-Member, to the grades of Student or Graduate for younger people up to about twenty-five or thirty years old.

In Britain it has always been possible for a boy on leaving school at fifteen to start work in the drawing office of a civil engineer, whether contractor or consultant, and eventually after many years of study in his spare time, to become a qualified civil engineer. This is becoming less easy and may soon become impossible. The recommended method of study for the ICE (Institution of Civil Engineers) examinations is now by full-time or sandwich study for a degree or diploma. Sandwich study is full-time work at a college interrupted by periods of full-time work with an employer.

Modern engineering requires more and more science, and to make use of its scientific theories, a civil engineer should study full-time for some years after leaving school. Therefore a university degree in civil engineering may soon become essential for membership of the ICE or any of the other civil engineering institutions (Institutions of Highway Engineers, Municipal Engineers, Public Health Engineers, Structural Engineers, Water Engineers, or the Permanent Way Institution, etc.).

To qualify for Associate-Membership of the ICE, a person must be at least twenty-six years old and working as a civil engineer. He must also pass certain examinations, satisfy ICE that he has had several years of useful engineering experience under the supervision of qualified civil engineers, both in the drawing office and on the site, and finally he must pass a mainly oral examination called the professional interview, before a group of qualified civil engineers. This is generally the only part of the examination from which candidates are never excused, whatever their civil engineering degree.

In general education, the minimum requirements, before a man may be accepted even as a candidate for the ICE examinations are as follows: five passes in the General Certificate of Education, (a) at advanced level in physics, (b) at advanced level in either pure or applied mathematics, (c) at ordinary level in English, and (d) at ordinary level in two other subjects. Detailed information is issued free by the ICE on all matters including the parts of the examination a candidate need not take as well as on the number of years and the types of civil engineering experience which are accepted.

In Britain the thirteen main engineering institutions were formally joined for examination purposes in 1965 in the Council of Engineering Institutions in London. A similar arrangement was made a few years earlier in the Unite Engineering Center, 345 East 47th Street, New York, for the United States institutions. In Britain all professions now take the Part 1 examination set by the Council of Engineering Institutions. This includes the five subjects of engineering drawing, mathematics, applied mechanics, principles of electricity, heat light and sound.

参考译文

第 1 课 土木工程

土木工程学，作为最古老的工程专业，是对建筑环境的规划、设计、施工和管理。这一环境包括从灌溉和排水系统到火箭发射设施的所有根据科学原理建造的建筑物。

土木工程师修筑道路、桥梁、隧道、大坝、海港、电厂、给排水系统、医院、学校、公交及其他现代社会和大量人口集中地所必需的基本公共设施；他们也修筑私有设施，诸如机场、铁路、管线、摩天大楼和其他为工业、商业或居住用途而设计的大型结构。此外，土木工程师还规划、设计和建造完整的城市与乡镇，最近已经开始规划和设计可容纳设备齐全的社区空间平台。

"土木"一词起源于拉丁语"citizen"。1782 年，英国人约翰·史密斯采用这一术语来区分他所从事的非军事工程的工作与在当时占主导地位的军事工程师的工程项目。从那时起，"土木工程"这个术语就经常用来指那些建造公共设施的工程师们，尽管该词所包含的范围更为广泛。

1. 范围

由于土木工程包含的范围如此广泛，所以它被细分为许多技术专业。根据工程类型的不同，需要多种土木工程专业人员。

当一个工程项目开始，土木工程师要进行场地的勘测和地形图的绘制，他们还要确定有效的给水、排水和电力线路的位置。岩土工程专家要进行土壤试验以确定土体是否能承受工程的荷载。环境工程专家要研究工程对当地环境的影响：对空气和地下水的潜在污染，对当地动植物的影响，以及工程如何设计才能满足政府针对环境保护的要求。交通工程专家要确定需要采用哪种设施来减轻因整个工程对当地道路和其他交通网造成的负担。同时，结构工程专家运用初始数据来对工程进行详细的设计、规划和说明。从项目开始到结束，监督和协调上述土木工程专家工作的是施工管理专家。根据其他专家提供的信息，施工管理专家估计所需的材料和人工的数量与造价，确定所有工作的进度，订购施工材料和设备，雇用承包商和分包商，并执行其他监督工作以确保工作按时且按规定完成。

在任何给定项目的整个过程中，土木工程师都大量运用计算机。计算机被用来设计和管理工程的多种要素（计算机辅助设计，或 CAD）。对现代土木工程师来说，计算机是个必需品，因为它们可以使工程师高效地处理大量的数据，而这些数据是确定工程的最佳建造方法所必需的。

2. 结构工程学

在这一专业领域，土木工程师规划和设计各种类型的结构，包括桥梁、大坝、电厂、设备支架、海上工程的特种结构、美国太空计划、发射塔、巨型天文和射电望远镜以及许多其他种类的工程。结构工程师用计算机确定结构所必须承受的力：自重、风（飓风）荷载、温度变化引起的建筑材料的膨胀和收缩，以及地震荷载。他们还需确定合适的材料组合，如钢、混凝土、塑料、石材、沥青、砖、铝或其他建筑材料。

3. 水力工程学

这个专业的土木工程师要处理水的物理控制的各个方面。他们的项目有助于预防洪水、为城市供水和灌溉用水、管理和控制河流和水径流、维护海滩及其他滨水设施。另外，他们还设计和维护港口、运河与水闸，建造大型水力发电大坝与小型坝及各种蓄水设施，帮助设计海域构筑物，并确定导航构筑物的位置。

4. 岩土工程学

专攻这一领域的土木工程师要分析支承结构并影响结构性能的岩土的性质。他们计算建筑和其他结构由于自重压力可能引起的沉降，并采取措施使之减少到最小。他们也评估和确定如何加强边坡和填充物的稳定性，以及如何保护结构免受地震和地下水的影响。

5. 环境工程学

在这一工程学分支中，土木工程师设计、建造和监督系统以提供安全的饮用水，防止和控制地面和地下供水的污染。他们还设计、建造和监督工程以控制或减轻对土壤和空气的污染。这些工程师建造供水和污水处理厂，设计空气净化器和其他设施把由于工业生产、焚化或其他排烟生产活动引起的空气污染减到最小或消除。他们还通过建造特殊倾倒场所或中和有毒有害物质来控制有毒有害废物的危害。此外，工程师还对垃圾掩埋进行设计和管理，以预防其对周围环境造成污染。

6. 交通工程学

从事这一专业的土木工程师建造设施以确保人和货物被安全高效地运输。他们精于设计和维护各种类型的交通设施，如高速公路和街道、公共交通系统、铁路和机场、港口和海港。交通工程师在设计每个项目中运用科技知识，并考虑经济、政治和社会方面的因素。他们与城市规划师紧密合作，因为社区的质量与交通系统的质量休戚相关。

7. 管道工程学

在土木工程学的这个分支中，工程师们建造管道和相关设施来输送液体、气体或固体，从煤浆（煤与水的混合）和半液态废物，到水、油和各种类型的高燃烧性和非燃烧性气体。工程师们决定管道设计，管道工程对它所必须通过地区的经济和环境的影响，所用材料类型——钢材、混凝土、塑料或多种材料的组合——的安装技术、测试管道强度的方法，以及控制所运送的流体材料保持适当的压力和流速。当输送有害物质时，安全也是一个主要的考虑因素。

8. 建造工程学

这一领域的土木工程师要从头到尾监督工程的施工。他们有时也被称为项目工程师，他们运用技术和管理技能，包括施工方法、规划、组织、财务以及运作施工项目方面的知识。事实上，他们协调参与工程的每一个人的行动：测量员，铺设和修筑临时道路和坡道、开挖基坑、支模和浇筑混凝土的工人，以及建造钢骨架的工人。这些工程师还要向工程业主做常规进度报告。

9. 社区和城市规划

从事这一领域工作的土木工程师为市内社区或整个城市进行规划和开发。此规划所包含的远不止工程方面，在使用和发展土地和自然资源中的环境、社会、经济因素也是重要因素。这些工程师要协调公共工程的规划和私人建筑发展之间的关系。他们评估所需的各类设施，包括街道和公路、公交系统、机场、港口设施、供水和废水处理系统、公共建筑、公园、休闲和其他保证社会、经济以及环境健康的设施。

10. 摄影、测量与绘图

该专业的土木工程师精确地测量地球表面以获得可靠信息来定位和设计工程项目。这一工作经常涉及高科技方法，诸如卫星和航空测量、摄影图像的计算机处理等。从卫星传来的无线电信号通过激光和声束扫描被转换成图形，为以下方面提供更精确的测量：隧道钻孔、修建高速公路和大坝、绘制洪水控制和灌溉工程图、定位可能影响建筑项目的地下岩石构成，以及其他许多建筑的使用。

11. 其他专业

另外两个工程专业虽不完全包含在土木工程的范围内，但对该学科很重要，它们就是工程管理和工程教育。

1）工程管理

许多工程师最终选择管理作为职业，而其他人可从管理这一位置开始他们的职业生涯。土木工程管理者综合了技术知识和组织协调劳动力、材料、机械和资金的能力。这些工程师可能在市、村镇、州或联邦政府工作；或在美国陆军工程兵团担任军事或民用管理工程师；或在半自治地区或城市权力机关或类似机构工作。他们也可能管理着规模由几个到数百人的私营工程公司。

2）工程教育

选择教育事业的土木工程师通常教授技术专业的研究生和本科生。许多土木工程教育者从事基础研究并引导建筑材料和施工方法的技术创新。许多人也担任工程项目顾问或与重要项目相关的技术部门或委员会的顾问。

Lesson 2
Basic Knowledge of Drawing[①]

1. Drawing tools and their utilization[②]

In order to improve the quality and efficiency in producing drawings, all tools must be used correctly.

1) Drawing board

A drawing board is used to fix drawing sheet and produce drawings on the sheet. The surface of the board should be flat and smoothy. The left side is the lead side and should be straight for guiding rulers as shown in Fig. 2-1.

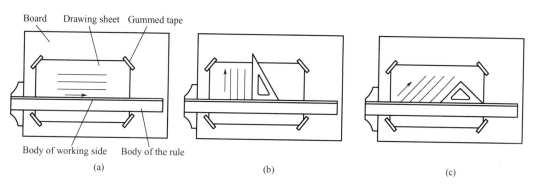

Fig. 2 - 1 The use of drawing board and T-square

(a) To fix paper and draw horizontal lines; (b) To draw vertical lines; (c) To draw parallel lines

2) T-square

A T-square is mainly used to draw horizontal lines. It can also be used to draw vertical lines in combination with a triangle as shown in Fig. 2 - 1(a). To draw horizontal lines, it needs to place the head of the T-square in contact to the left edge of the drawing board with left hand and move the T-square to the desired position. Then it needs to hold the pencil and draw the line from its left end to the right end. One should not directly use a T-square to draw vertical lines and should not use the lower edge of a T-square to draw horizontal lines either.

3) Triangles

A triangle is often used in combination with a T-square to draw vertical lines, lines with an inclination angle of 30°, 45° or 60° to horizontal lines as shown in Fig. 2 - 1(b) and (c). One can also use two triangles to draw parallel lines and perpendicular lines of any orientations.

    4) Other drawing tools

(1) Compass.

A compass is used to draw circles and arcs. The leg of the compass with a step pin should be facing downward and the pencil tip on the other leg should have similar height to the pin as shown in Fig. 2-2(a). When drawing on paper, one can rotate the compass by revolving the handle clockwise and incline the compass slightly forward as shown in Fig. 2-2(b). When drawing circles with different diameters, the pin and the pencil of the compass should be adjusted to make them perpendicular to the paper as shown in Fig. 2-2(c) and Fig. 2-2(d).

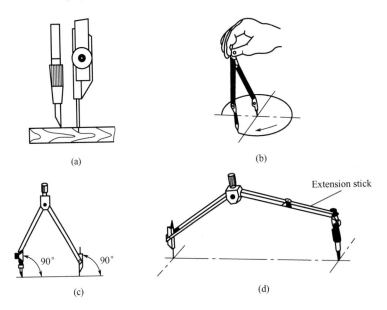

Fig. 2-2 The use of a compass

(a) Needle point should be slightly longer in the pencil end;

(b) To draw a circle in clockwise direction;

(c) Steel needle and pencil lead straddle are perpendicular to paper;

(d) To add an extension stick to draw big circles

(2) Dividers.

Dividers are used for transferring distances and for equal subdivision of lines and circles. The two tips should meet together at the position shown in Fig. 2-3.

2. Related provisions in national standards

Engineering drawing is an important documentation used during the process of design and construction. In order to direct production and technical communication, ensure the quality of drawing, improve the effect of drawing, and meet the requirements of design, construction and archiving, all drawings should comply with related provisions in national standards. National standards are abbreviated as "GB".

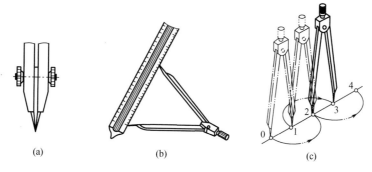

Fig. 2-3   The use of dividers

(a) Put needle tips together; (b) Measure the size; (c) Size with equal subdivision

1) Sheet size and format of drawing (GB 50001—2010)

(1) Sheet size.

While producing the engineering drawings, one should adopt standard basic size illustrated in Table 2-1. There are five standard sheet sizes, namely A0, A1, A2, A3 and A4. If it is necessary, one may also use extended sheets of larger dimensions specified in the standard.

Table 2-1   Basic sheet size (mm)

Dimensions \ Size code	A0	A1	A2	A3	A4
$b \times l$	841×1189	594×841	420×594	297×420	210×297
c	10			5	
a	25				

(2) Frame format.

One must use continuous thick lines to draw the borders (Fig. 2-4). Drawing sheets can be placed in both horizontal type and vertical type. Generally, for A0—A3 drawings sheets one should adopt horizontal type, and can also adopt vertical type when in necessary.

(3) Title block and signature block.

One must draw a title block on each drawing sheet. The position of the title block should be located on the lower right corner of the sheet as shown in Fig. 2-5(a). The size, form or partition of the title block should be determined according to the engineering needs, and the simplified title block can be used in the college assignments [Fig. 2-5(b)].

(4) Setting order of sheets.

Engineering sheets should be arranged in order of specialty, that is drawing content, general layout, architectural drawings, structural drawings, water supply and sewerage drawings, heating and air-conditioning drawings, electric drawings, and so on. Different professional drawings should be ordered in the primary-secondary relation or logical

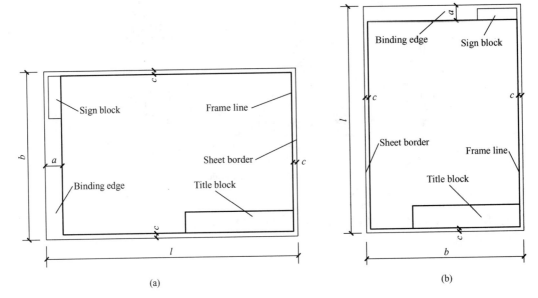

Fig. 2 - 4　Forms of drawing sheets

(a) Horizontal type of A0—A3 drawing sheets；(b) Vertical type of A0—A3 drawing sheets

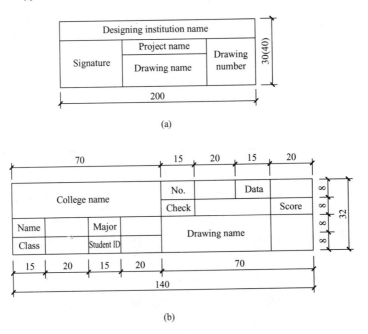

Fig. 2 - 5　Forms of title block

(a) Title block；(b) Title block of assignment

relation of drawings.

2) Scales

The ratio between the dimension on the drawing and features of the actual object is called the scale of the drawing. One should use an appropriate scale as suggested in Table 2 - 2. In general, the same drawing should choose either a kind of scale or two kinds of scales according to the needs of professional drawings. In a particular case, one can choose scales by oneself but it needs to draw the relevant scale in the appropriate location in addition to the marked drawing scale.

Table 2-2 Scales

Common scales	1:1, 1:2, 1:5, 1:10, 1:20, 1:50, 1:100, 1:150, 1:200, 1:500, 1:1000, 1:2000, 1:5000, 1:10000, 1:20000, 1:50000, 1:100000, 1:200000
Usable scales	1:3, 1:4, 1:6, 1:15, 1:25, 1:40, 1:60, 1:80, 1:250, 1:300, 1:400, 1:600

3) Lettering

Lettering of characters, numbers or symbols on a drawing must follow related national standards:

(1) The characters must be whole and clear. The distance between characters must be uniformly distributed, standing in a line.

(2) The font for Chinese characters should be the "long Simsong" and the standard simplified Chinese characters officially issued by the Chinese government should be used. The height of characters should not be less than 3.5mm while the basic height of characters is 3.5mm, 5mm, 7mm, 10mm, 14mm, 20mm.

(3) The letters and numbers can be written in italic font or normal/straight font. The height of letters and numbers cannot be smaller than 2.5mm, and the width should be about one tenth or one fourteenth of the character height. The character in italic font should be inclined towards the right with an angel of 75° with respect to horizontal lines. In general, italic font is used for all types of engineering drawings.

(4) The letters and numbers for exponents, fractions, the limit deviation, notes, etc. are usually used by one grade less than the full size fonts in the drawing.

Several examples are shown in Fig. 2-6.

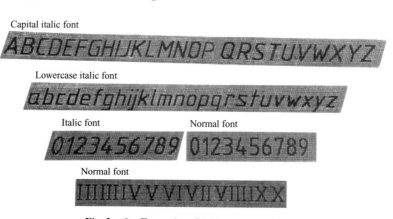

Fig. 2-6 Examples of letters and numbers

4) Line styles

All lines names, styles, thicknesses and their applications are defined in the national standard. Line styles are shown in Table 2 – 3. When producing drawings, the width of lines should be chosen in 0.35mm, 0.5mm, 0.7mm, 1.0mm, 1.4mm or 2.0mm. Line width group is shown in Table 2 – 4. The width scale of thick lines and thin lines is 2 : 1, and the width of thick lines is chosen on the basis of the complex degree and scale of engineering drawings.

Table 2 – 3　Line types

Description		Line types	Width	Usage
Continuous line	thick		b	Main visible outlines
	half-width		$0.5b$	Visible out lines, etc.
	thin		$0.25b$	Dimension lines, etc.
Long Dashed line	thick		b	See drawing standards of each related specialty
	half-width		$0.5b$	Hidden outlines
	thin		$0.25b$	Hidden outlines, materials example lines, etc.
Long Dashed dotted line	thick		b	See drawing standards of each related specialty
	half-width		$0.5b$	See drawing standards of each related specialty
	thin		$0.25b$	Centre lines, symmetric lines, axial lines, etc.
Straight line with intermittent zigzags	thin		$0.25b$	Boundary lines for broken views
Irregular line	thin		$0.25b$	Boundary lines for broken views

Table 2 – 4　Line width group（mm）

Ratio of line width	Line width group					
b	2.0	1.4	1.0	0.7	0.5	0.35
$0.5b$	1.0	0.7	0.5	0.35	0.25	0.18
$0.25b$	0.5	0.35	0.25	0.18	—	—

Lesson 2 Basic Knowledge of Drawing

5) Dimensioning

A complete dimension, as shown in Fig. 2 – 7, is composed of an extension line, a dimension line, start-stop symbols and dimension text.

(1) Extension line.

An extension line drawn in a continuous thin line illustrates the scope of the corresponding dimension, as shown in Fig. 2 – 8. An extension line is originated from a feature outline, an axis line, or a symmetrical center line. The outline, axis line and symmetrical center line can also be used as an extension line.

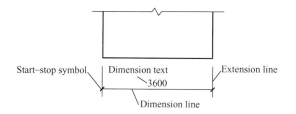
Fig. 2 – 7 Dimension construction

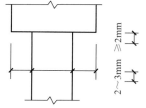
Fig. 2 – 8 Extension line

An extension line is usually drawn perpendicular to the dimension lines, which one end should leave away from outline of drawing no less than 2mm and the other end should extend beyond the terminal of dimension line for about 2 — 3mm. If necessarily, the extension line may also have a different inclination angle with the dimension line.

(2) Dimension line.

A dimension line must be drawn in a continuous thin line, neither be replaced by existing lines in the drawing nor be in coincidence with any existing lines or drawn as extension of any existing lines.

(3) Start-stop symbol.

A start-stop symbol usually is drawn in the inclined short half-width line, and its inclined angle should be 45° angle in clockwise and the length should be 2 — 3mm [Fig. 2 – 9(a)]. The start-stop symbols of radius, diameter, angle and arc length should be shown in arrowheads [Fig. 2 – 9(b)].

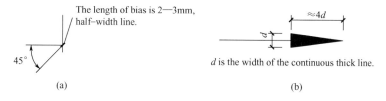
Fig. 2 – 9 Kinds of dimension line terminal
(a) Bias; (b) Arrow

(4) Dimension text.

Dimension text should usually be marked above the dimension line or in the breaking space of a dimension line. It should, however, be kept consistent in the same drawing. When there is not enough space, it can also be placed somewhere else by using a note. As shown in Fig. 2-10, all linear dimensions should usually be marked heading upwards for horizontal dimensions, heading towards the left for vertical dimensions, or with an inclination heading upwards. Dimension text should be marked outside outline of drawing and should not intersect the drawing lines, texts, symbols, and so on.

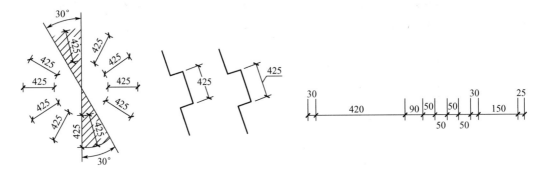

Fig. 2-10 Writing direction and location of dimension text

For linear dimension, the dimension line must be parallel to the line segment being dimensioned. The distance between the dimension line and the outmost feature should not be less than 10mm. All distances between neighboring parallel dimension lines should be as consistent as possible and are usually with 7—10mm separation. A small size is usually placed inside and a big size is usually placed outside, as shown in Fig. 2-11.

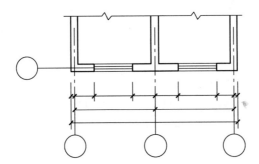

Fig. 2-11 Arrangement of dimension lines

I. New Words

1. horizontal *adj.* 水平的，地平线的；*n.* 水平线，水平面
2. vertical *adj.* 垂直的，直立的；*n.* 垂直线，垂直面
3. triangle *n.* 三角板，三角（形）
4. inclination *n.* 倾斜，倾向，斜坡
5. parallel *adj.* 平行的；*vt.* 使……与……平行；*n.* 平行线
6. perpendicular *adj.* 垂直的，正交的，直立的；*n.* 垂线
7. orientation *n.* 方向，定向，向东方

Lesson 2 Basic Knowledge of Drawing

8. compass *n.* 圆规，指南针，罗盘；*vt.* 包围
9. incline *vt.* 使倾斜，使倾向于；*vi.* 倾斜，倾向；*n.* 倾斜，斜面，斜坡
10. divider *n.* 圆规，分隔物
11. subdivision *n.* 分部，细分
12. documentation *n.* 文件，证明文件，文件编制
13. construction *n.* 施工，建筑物，构造
14. logical *adj.* 合逻辑的，合理的，逻辑学的
15. symmetrical *adj.* 匀称的，对称的
16. coincidence *n.* 一致，同时发生
17. intersect *vt.* & *vi.* 相交，交叉，横断，贯穿

Ⅱ. Phrases and Expressions

1. drawing board   图板
2. drawing sheet   图纸
3. T-square   丁字尺
4. sheet size   图幅
5. title block   标题栏
6. signature block   会签栏
7. architectural drawing   建筑图
8. water supply and sewerage   给排水
9. continuous line   实线
10. dashed line   虚线
11. dashed dotted line   点画线
12. extension line   尺寸界线
13. dimension line   尺寸线
14. start-stop symbol   起止符号
15. dimension text   尺寸数字
16. linear dimension   线性尺寸

Ⅲ. Notes

① 本篇文章的内容虽然比较简单，但相关的专业英语词汇或短语较多，这也是科技英语区别于普通英语的词汇特征。科技英语词汇基本为书面用语，且意义比较专一、稳定，要求不能引起歧义，表达方式也较简单，语汇不具感情色彩，不追求词形悦目与语音悦耳，因此很少使用文学英语中常用的比喻、排比、夸张等修辞手段。

② 本篇文章具有鲜明的科技英语的语法特点，即不仅大量使用被动语态与一般现在时，而且大量使用非谓语动词短语等，尤其是标题，大量使用名词或名词词组，简单明了。

Ⅳ. Exercises

Fill in the blanks with the information given in the text.

1. A T-square can be directly used to draw _____ lines. It can also be used to draw _____ lines in combination with a triangle.

2. The position of the _____ should be located on the lower right corner of the sheet.

3. The ratio between the dimension on the drawing and features of the actual object is called _____ of the drawing.

4. A complete dimension is composed of _____ and dimension text.

Ⅴ. Expanding

Remember the following terms related to the engineering drawing.

1. graphics　制图学
2. descriptive geometry　画法几何学
3. location axis grid　定位轴线
4. projection method　投影法
5. central projection method　中心投影法
6. parallel projection method　平行投影法
7. orthogonal projection method　正投影法
8. oblique projection method　斜投影法
9. project center　投射中心
10. view　视图
11. plane drawing　平面图
12. elevation drawing　立面图
13. section drawing　剖面图
14. cross-section drawing　横截面图
15. axonometric drawing　轴测图
16. perspective drawing　透视图

Ⅵ. Reading Material

Principles of Orthographic Projection

One of the important techniques for producing engineering drawings is orthographic projection. Orthographic projection forms the basis for producing and reading engineering drawings.

1. Form of projection

In daily life, when the sunlight or lamplight irradiates an object, there will be a shadow of the object on the ground or wall. People abstract the natural phenomena with science to describe an object with projection. The source of light is called the center of projection. The light ray is called the line of projection. The preestablished plane is called the plane of projection. The graphics on the plane is called the projection of the object (Fig. 2-12). When the line of projection passes through the object and the preestablished plane, there will be graphics produced on the plane. This method is called projection.

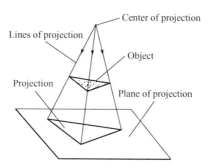

Fig. 2-12 Central projection

2. Types of projection

There are two types of projection, central projection and parallel projection.

1) Central projection

When the lines of projection meet at a point, it is called central projection (Fig. 2-12).

The size of the projected drawing using central projection depends on the distance between the object and the plane of projection. It can show the exact dimension of the object and thus the central projection is mainly used for the pictorial drawings of buildings.

2) Parallel projection

When the lines of projection are parallel with each other, it is called parallel projection (Fig. 2-13). Parallel projection includes oblique projection and orthographic projection.

Oblique projection: When the lines of projection are not orthogonal to the plane of projection, it is called oblique projection [Fig. 2-13(a)].

Orthographic projection: When the lines of projection are orthogonal to the plane of projection, it is called orthographic projection [Fig. 2-13(b)].

Because orthographic projection can reflect the true size of the object, it is used in most engineering drawings. For convenience of description, "orthographic projection" is called projection for short later.

3. Features of orthographic projection

1) Authenticity

When planar features or lines are in parallel with the projection plane, the projection shows the real shape or true length of the features being projected [Fig. 2-14(a)].

2) Accumulation

When planar features or lines are orthogonal to the projection plane, the projection

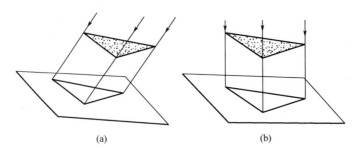

Fig. 2 – 13 Parallel projection

(a) Oblique projection; (b) Orthographic projection

accumulates to a line or a point [Fig. 2 – 14(b)].

3) Similarity

When planar features or lines are inclined to the projection plane in general, the projection produces similar shapes of the original feature. The length of projected lines will be shorter than the true length [Fig. 2 – 14(c)]. In similar shapes, the parallelism of lines before and after projection does not change and the connectivity of features does not change either after projection.

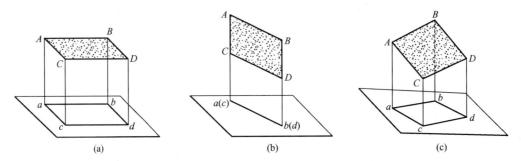

Fig. 2 – 14 Characteristics of orthographic projection

(a) Authenticity; (b) Accumulation; (c) Similarity

4. Three-view drawings

A drawing produced through orthographic projection is called a view of the object. Because one view can only show the shape of the object from two sides, it cannot represent the whole object. It is necessary to use multiple views in engineering drawings.

First, put one object in a projection system with three projection planes that are orthogonal from each other. The object should be placed between the viewer and the projection planes, respectively. Orthographic projection is used for producing three-view projections. In the three-view projection system, the three projection planes are orthogonal from each other and called H plane, V plane and W plane, respectively. The H plane is a horizontal projection plane and is called H-plane in short. The V plane is a frontal vertical plane and is called V-plane in short. The W plane is a profile plane and is called W-plane in short. The intersection lines of the three projection planes are called the axes of projection, which are termed as X-axis, Y-axis and Z-axis respectively. These axes meet at a point which is called the origin. The drawing produced on the frontal plane (V-plane) is called orthographic drawing, also called front view. The drawing produced on the horizontal projection plane (H-plane) is called horizontal projection, also called top view. The drawing on the profile plane (W-plane) is called profile projection, also called left-side view [Fig. 2-15(a)].

To draw three-view drawings on a drawing sheet, following the national standard (GB), the V-plane is first fixed. The H-plane turns down 90° about the OX-axis along the arrow direction and the W-plane turns right 90° about the OZ-axis as shown in Fig. 2-15(b). As a result, the H-plane and the W-plane coincide with the V-plane and the front view, the top view and the left-side view are on the same plane [Fig. 2-15(c)]. Since the size of the projection planes is not related to the views, it is not necessary to draw the border of the projection planes and the distance between individual views is decided by the breadth of the drawing sheet and the size of the views [Fig. 2-15(d)].

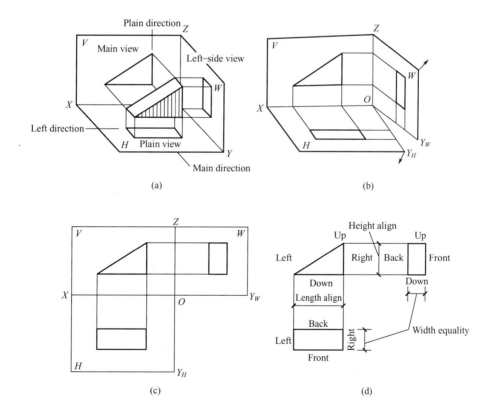

Fig. 2-15 Form of three-view drawings

(a) Object in a projection system with three projection planes;

(b) The system with three projection planes will be flattening;

(c) The relationship between the locations of three-views;

(d) Three-view drawings of the "three equal" and the relationship between the position

参 考 译 文

第 2 课　制图基本知识

1. 制图工具及其用法

为了提高绘图质量和效率,必须正确地使用绘图工具。

1) 图板

图板是用来固定图纸并进行绘图的。板面要求平整光滑,左侧为导边,必须平直,如图 2-1 所示。

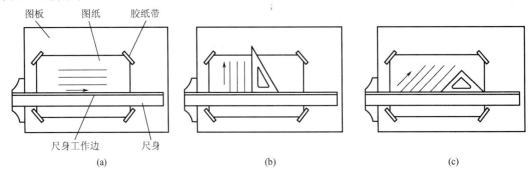

图 2-1　图板及丁字尺的应用

(a) 固定图纸及画水平线;(b) 画竖直线;(c) 画平行线

2) 丁字尺

丁字尺主要用来画水平线,还常与三角板配合画铅垂线,如图 2-1(a) 所示。画水平线需用左手扶住尺头紧靠图板左侧导边,上下滑移到所需位置。然后拿着铅笔沿丁字尺工作边自左向右画水平线。禁止直接用丁字尺画铅垂线,也不能用尺身下缘画水平线。

3) 三角板

三角板常与丁字尺配合使用,画水平线的垂直线,以及与水平成 30°、45°或 60°的斜线,如图 2-1(b) 和 (c) 所示。两块三角板配合使用,可画任意方向倾斜线的平行线和垂直线。

4) 其他绘图工具

(1) 圆规。

圆规用来画圆和圆弧。使用时,应将圆规钢针有台阶的一端朝下,并使台阶面与铅芯平齐,如图 2-2(a) 所示。画图时,按顺时针方向旋转并稍向前倾斜,如图 2-2(b) 所示。画不同直径的圆时,要注意随时调整钢针和铅芯插腿,使其始终垂直于纸面,如图 2-2(c) 和图 2-2(d) 所示。

(2) 分规。

分规用来量取和等分线段和圆。当两脚并拢时,两针尖应对齐,其用法如图 2-3 所示。

2. 国家标准有关规定

工程图样是设计和施工过程中的重要技术资料。为了便于指导生产和技术交流,保证

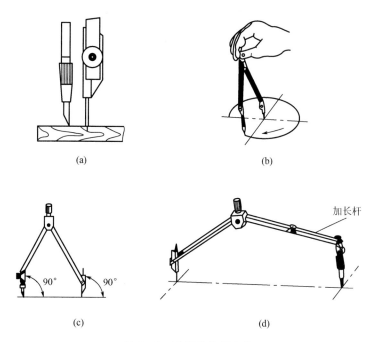

图 2-2 圆规的使用方法

(a) 针尖略长于铅芯；(b) 按顺时针方向画圆；
(c) 钢针和铅芯插腿垂直于纸面；(d) 用加长杆画大圆

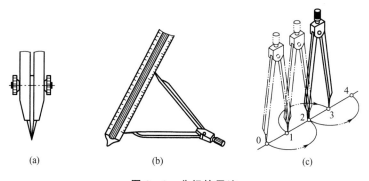

图 2-3 分规的用法

(a) 针尖并拢于一点；(b) 量取尺寸；(c) 连续截取等长线段

制图质量，提高制图效果，符合设计、施工、存档的要求，所有工程图样必须符合国家标准的有关规定。国家标准简称"国标"，用符号"GB"表示。

1) 图纸幅面和格式（GB 50001—2010）

(1) 图纸幅面。

绘制图样时，应采用表 2-1 中规定的基本幅面。有五种标准幅面，即 A0、A1、A2、A3 及 A4 幅面。必要时也可以选用国标规定的更大尺寸的加长幅面。

表 2-1　基本图幅 (mm)

尺寸 \ 幅面代号	A0	A1	A2	A3	A4
$b×l$	841×1189	594×841	420×594	297×420	210×297
c	10			5	
a	25				

(2) 图框格式。

在图纸上必须用粗实线画出图框（图 2-4）。图纸有横式和立式两种。一般 A0～A3 图纸宜采用横式，必要时也可采用立式。

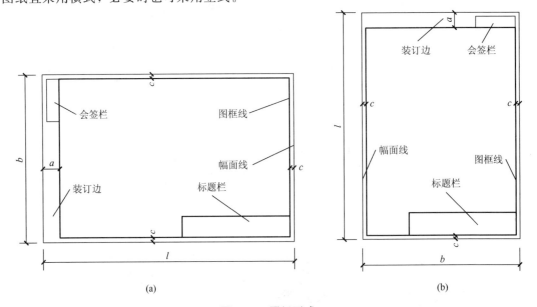

图 2-4　图纸形式
(a) A0～A3 横式幅面；(b) A0～A3 立式幅面

(3) 标题栏与会签栏。

每张图纸上都必须画出标题栏。标题栏应位于图纸的右下角，如图 2-5(a) 所示。标题栏应根据工程需要确定其尺寸、格式及分区，学校的制图作业可采用简化的标题栏[图 2-5(b)]。

(4) 图纸编排顺序。

工程图纸应按照专业顺序编排，应为图纸目录、总图、建筑图、结构图、给水排水图、暖通空调图、电气图等。各专业的图纸应按图纸内容的主次关系、逻辑关系进行排序。

2) 比例

图中图形与实物相对应的线性尺寸之比，称为比例。绘图时，可从表 2-2 规定的系列中选取适当的比例。一般情况下，一个图样应选择一种比例，根据专业制图需要，同一图样可选用两种比例。特殊情况下，也可自选比例，这时除了标注绘图比例外，还应在适

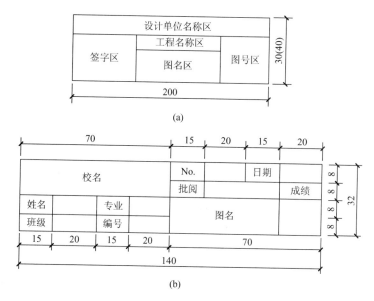

图 2-5 标题栏格式

（a）标题栏；（b）作业标题栏

当位置绘制出相应的比例尺。

表 2-2 绘图的比例

常用比例	1:1、1:2、1:5、1:10、1:20、1:50、1:100、1:150、1:200、1:500、1:1000、1:2000、1:5000、1:10000、1:20000、1:50000、1:100000、1:200000
可用比例	1:3、1:4、1:6、1:15、1:25、1:40、1:60、1:80、1:250、1:300、1:400、1:600

3）字体

图纸上所书写的文字、数字或符号，必须按国标规定书写，应做到：

（1）字体工整、笔画清楚、间隔均匀、排列整齐。

（2）汉字应写成长仿宋字，并采用国家正式颁布推行的简化字。字高度不应小于 3.5mm，其基本字高尺寸为：3.5mm、5mm、7mm、10mm、14mm、20mm。

（3）字母和数字可写成斜体或正体。字母和数字的字高不应小于 2.5mm，其宽度约为字高的 1/10 或 1/14。斜体字的字头向右倾斜，与水平线成 75°角。图样上一般采用斜体字。

（4）用作指数、分数、极限偏差、注脚等的字母及数字，一般采用小一号字体。

字体示例如图 2-6 所示。

4）图线

国标规定了各种图线的名称、形式、宽度及其应用，线型如表 2-3 所示。图线的宽度 b 宜从 0.35mm、0.5mm、0.7mm、1.0mm、1.4mm、2.0mm 的线宽系列中选取，线宽组如表 2-4 所示。粗线与细线宽度之比为 2:1，绘图时粗线宽度应根据工程图的复杂程度与比例大小确定。

图 2-6 字母和数字字体示例

表 2-3 图线形式

名　　称		线　　型	线宽	应　　用
实线	粗	——————	b	主要可见轮廓线
	中	——————	0.5b	可见轮廓线等
	细	——————	0.25b	尺寸线等
虚线	粗	— — — —	b	见各有关专业制图标准
	中	- - - - - -	0.5b	不可见轮廓线
	细	- - - - - - -	0.25b	不可见轮廓线、图例线等
单点长画线	粗	—·—·—	b	见各有关专业制图标准
	中	—·—·—	0.5b	见各有关专业制图标准
	细	—·—·—	0.25b	中心线、对称线、轴线等
折断线	细	——∿——	0.25b	断开界线
波浪线	细	～～～	0.25b	断开界线

表 2-4 线宽组（mm）

线 宽 比	线　宽　组					
b	2.0	1.4	1.0	0.7	0.5	0.35
0.5b	1.0	0.7	0.5	0.35	0.25	0.18
0.25b	0.5	0.35	0.25	0.18	—	—

5) 尺寸注法

一个完整的尺寸由尺寸界线、尺寸线、尺寸起止符号以及尺寸数字组成，如图 2-7 所示。

(1) 尺寸界线。

尺寸界线用细实线画出，表示所注尺寸的界限范围，如图 2-8 所示。尺寸界线应由图形的轮廓线、轴线或对称中心线处引出，也可利用轮廓线、轴线或对称中心线作尺寸界线。

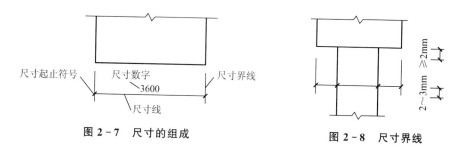

图 2-7 尺寸的组成　　　　图 2-8 尺寸界线

尺寸界线一般与尺寸线垂直，一端离开图样轮廓线不应小于 2mm，另一端超出尺寸线 2～3mm。必要时也允许与尺寸线倾斜。

（2）尺寸线。

尺寸线必须用细实线绘制，不能用图上任何其他图线代替，也不能与图线重合或画在图线的延长线上。

（3）尺寸起止符号。

尺寸起止符号一般用中粗斜短线绘制，其倾斜方向应与尺寸界线成顺时针 45°角，长度宜为 2～3mm [图 2-9(a)]。半径、直径、角度与弧长的尺寸起止符号宜用箭头表示 [图 2-9(b)]。

图 2-9 尺寸线终端的形式
（a）斜线；（b）箭头

（4）尺寸数字。

线性尺寸的尺寸数字一般应注写在尺寸线的上方或中断处，但同一图样上最好保持一致。当位置不够时，也可引出标注。如图 2-10 所示，线性尺寸数字一般按水平方向字头朝上方向注写，垂直方向字头朝左，倾斜方向字头朝上书写。尺寸宜标注在图样轮廓线以外，不宜与图线、文字及符号等相交。

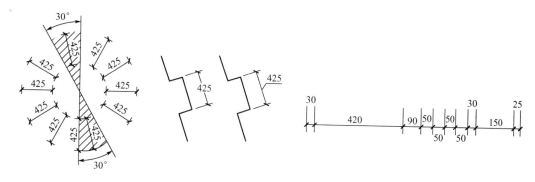

图 2-10 尺寸数字的注写方向与位置

标注线性尺寸时，尺寸线必须与所标注的线段平行。尺寸线与最外边轮廓线的间隔不宜小于 10mm。当几条尺寸线互相平行时，其间隔尽量保持一致，一般间隔为 7～10mm，且小尺寸在内，大尺寸在外，如图 2-11 所示。

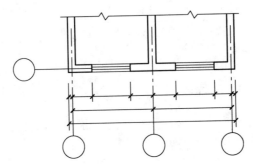

图 2-11　尺寸线布置

Lesson 3
Building and Building System

1. The building

Architecture and building construction are not necessarily one and the same thing. An understanding of the methods for assembling various materials, elements, and components is necessary during both the design and the construction of a building. <u>This understanding, however, while it enables one to build architecture, does not guarantee it.</u>① A working knowledge of building construction is only one of several critical factors in the execution of architecture. When we speak of architecture as the art of building, <u>we should consider the following conceptual systems of order in addition to the physical ones of construction</u>②:

(1) The definition, scale, proportion, and organization of the interior spaces of abuilding.

(2) The ordering of human activities by their scale and dimension.

(3) The functional zoning of the spaces of a building according to purpose and use.

(4) Access to the horizontal and vertical paths of movement through the interior of a building.

(5) The sensible qualities of a building: form, space, light, color, texture, and pattern.

(6) The building as an integrated component within the natural and built environment.

<u>Of primary interest to us in this text are the physical systems that define, organize, and reinforce the perceptual and conceptual ordering of a building.</u>③

A system can be defined as an assembly of interrelated or interdependent parts forming a more complex and unified whole and serving a common purpose. A building can be understood to be the physical embodiment of a number of systems and subsystems that must necessarily be related, coordinated, and integrated with each other as well as with the three-dimensional form and spatial organization of the building as a whole (Fig. 3 – 1).

2. Building system

1) Structural system (Fig. 3 – 2)

The structural system of a building is designed and constructed to support and transmit applied gravity and lateral loads safely to the ground without exceeding the allowable stresses in its members.

(1) The superstructure is the vertical extension of a building above the foundation.

(2) Columns, beams, and load-bearing walls support floor and roof structures.

(3) The substructure is the underlying structure forming the foundation of a building.

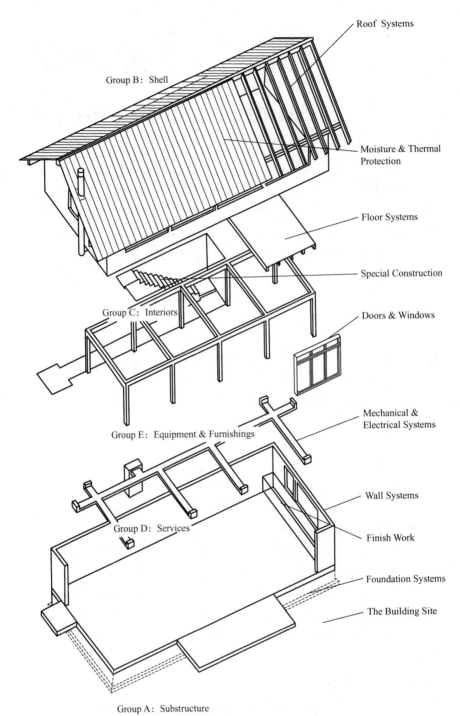

Fig. 3 – 1　Building system

2) Enclosure system

The enclosure system is the shell or envelope of a building, consisting of the roof, exterior walls, windows, and doors.

Lesson 3 Building and Building System

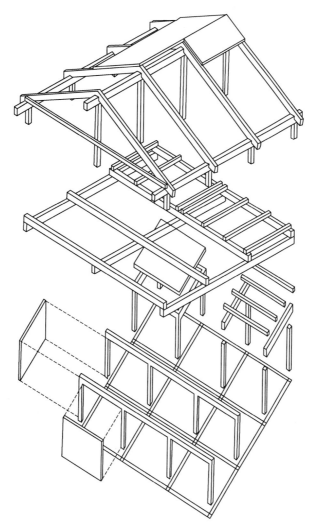

Fig. 3-2 Structural system

(1) The roof and exterior walls shelter interior spaces from inclement weather and control moisture, heat, and air flow through the layering of construction assemblies.

(2) Exterior walls and roofs also dampen noise and provide security and privacy for the occupants of a building.

(3) Doors provide physical access.

(4) Windows provide access to light, air, and views.

(5) Interior walls and partitions subdivide the interior of a building into spatial units.

3) Mechanical systems

The mechanical systems of a building provide essential services to a building.

(1) The water supply system provides potable water for human consumption and sanitation.

(2) The sewage disposal system removes fluid waste and organic matter from a building.

(3) Heating, ventilating, and air-conditioning systems condition the interior spaces of

a building for the environmental comfort of the occupants.

(4) The electrical system controls, meters, and protects the electric power supply to a building, and distributes it in a safe manner for power, lighting, security, and communication systems.

(5) Vertical transportation systems carry people and goods from one level to another in medium-and high-rise buildings.

(6) Fire-fighting systems detect and extinguish fires.

(7) Structures may also require waste disposal and recycling systems.

3. Considerations of building systems in construction

The manner in which we select, assemble, and integrate the various building systems in construction should take into account the following factors:

1) Performance requirements

(1) Structural compatibility, integration, and safety.

(2) Fire resistance, prevention, and safety.

(3) Allowable or desirable thickness of construction assemblies.

(4) Control of heat and air flow through building assemblies.

(5) Control of migration and condensation of water vapor.

(6) Accommodation of building movement due to settlement, structural deflection, and expansion or contraction with changes in temperature and humidity.

(7) Noise reduction, sound isolation, and acoustical privacy.

(8) Resistance to wear, corrosion, and weathering.

(9) Finish, cleanliness, and maintenance requirements.

(10) Safety in use.

2) Aesthetic qualities

(1) Desired relationship of building to its site, adjacent properties, and neighborhood.

(2) Preferred qualities of form, massing, color, pattern, texture, and detail.

3) Regulatory constraints

Compliance with zoning ordinances and building codes.

4) Economic considerations

(1) Initial cost comprising material, transportation, equipment, and labor costs.

(2) Life-cycle costs, which include not only initial cost, but also maintenance and operating costs, energy consumption, useful lifetime, demolition and replacement costs, and interest on invested money.

5) Environmental impact

(1) Conservation of energy and resources through siting and building design.

(2) Energy efficiency of mechanical systems.

(3) Use of resource-efficient and nontoxic materials.

6) Construction practices

(1) Safety requirements.

(2) Allowable tolerances and appropriate fit.

(3) Conformance to industry standards and assurance.

(4) Division of work between the shop and the field.

(5) Division of labor and coordination of building trades.

(6) Budget constraints.

(7) Construction equipment required.

(8) Erection time required.

(9) Provisions for inclement weather.

Ⅰ. New Words

1. component *n.* 成分，组件，元件
2. guarantee *n.* 保证，担保，保证人；*vt.* 保证，担保
3. execution *n.* 执行，实行，完成
4. conceptual *adj.* 概念上的，观念的
5. functional *adj.* 功能的，机能的
6. horizontal *adj.* 水平的，地平线的
7. vertical *adj.* 垂直的，直立的
8. texture *n.* 质地，纹理，本质
9. integrate *vt.* 使……成整体，求……的积分；*adj.* 整合的，完全的
10. perceptual *adj.* 感知的，有知觉的
11. unify *vt.* 统一，使一致，使相同
12. embodiment *n.* 体现，化身，具体化
13. subsystem *n.* 子系统，次要系统
14. spatial *adj.* 空间的，受空间限制的
15. shelter *n.* 庇护所，避难所，遮盖物
16. inclement *adj.* 险恶的，气候严酷的
17. dampen *vt.* 抑制，使……沮丧，使……潮湿
18. detect *vt.* 察觉，发现，探测
19. extinguish *vt.* 熄灭，压制，偿清
20. compatibility *n.* 兼容性，适用性
21. integration *n.* 综合，集成
22. migration *n.* 迁移，移民，移动
23. accommodation *n.* 住处，膳宿，调节
24. deflection *n.* 挠曲，偏差，偏向
25. acoustical *adj.* 声学的，听觉的，音响的
26. aesthetic *adj.* 美的，美学的
27. regulatory *adj.* 管理的，控制的，调整的
28. constraint *n.* 约束，限制
29. demolition *n.* 拆除（等于 demolishment），拆迁，破坏
30. tolerance *n.* 公差，宽容，容忍
31. conformance *n.* 一致，适合，顺从
32. budget *n.* 预算，预算费；*vt.* 编预算，安排

Ⅱ. Phrases and Expressions

1. load-bearing wall 承重墙
2. sewage disposal system 污水处理系统
3. mechanical system 机械系统
4. electric power supply 电力供应
5. vertical transportation system 垂直交通系统
6. high-rise building 高层建筑
7. fire-fighting system 消防系统
8. environmental impact 环境影响

Ⅲ. Notes

① 句中第一个"it"指的是"this understanding"；第二个"it"指的是"build architecture"，"guarantee it"指的是保证施工过程中的质量。

② 句中的"ones"指代"systems"，"the physical ones"指代实际施工体系。

③ 此句采用倒装句，由于主语"the physical systems"带有定语从句较长，谓语较短，为避免整个句子显得头重脚轻而放在后面。

Ⅳ. Exercises

Translate the professional terms in Fig. 3-3 and Fig. 3-4 into English with the help of a dictionary.

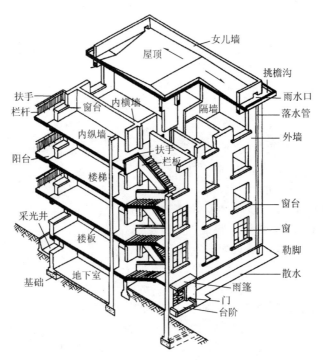

Fig. 3-3 砖混结构建筑的组成
Components of a brick and concrete composite structure building

Lesson 3 Building and Building System

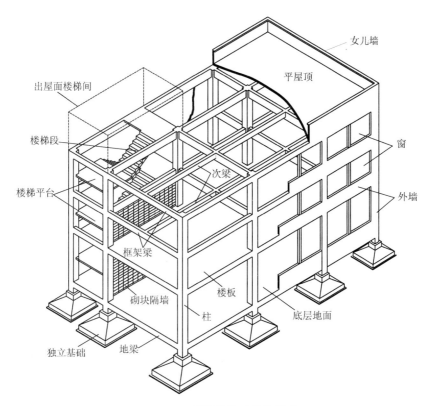

Fig. 3 – 4 框架结构建筑的组成
Components of a frame structure building

V. Expanding

Remember the following terms related to the building and architectural design.

1. architectural style   建筑形式
2. architectural function/building function   建筑功能
3. architectural image   建筑形象
4. civil building   民用建筑
5. industrial building   工业建筑
6. agricultural building   农业建筑
7. designed service life   设计使用年限
8. fire resistance rating/fire resistance classification   耐火等级
9. fire endurance   耐火极限
10. combustibility   燃烧性能
11. architectural design   建筑设计
12. structure design   结构设计
13. equipment design   设备设计
14. general plan/master plan/overall plan   总体规划

Ⅵ. Reading Material

Reinforced Concrete Members

Reinforced concrete structures consist of a series of individual "members" that interact to support the loads placed on the structure. Although these are considered separately in the design process, their interaction must also be taken into account. This is done in the overall analysis of the structure and in the design of joints and connections. The more important types of elements are illustrated in Fig. 3−5 and Fig. 3−6.

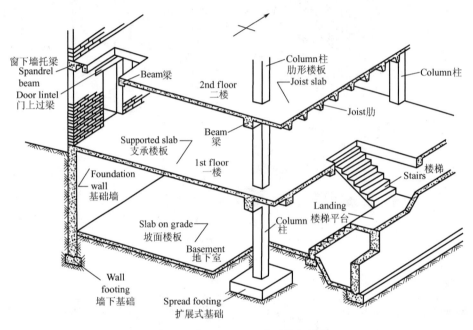

Fig. 3−5 Reinforced concrete building elements

The second floor of the building in Fig. 3−5 is built of concrete joist-slab construction. Here a series of parallel ribs or joists support the load from the top slab. The reactions supporting the joists apply loads to the beams, which in turn are supported by columns. In such a floor, the top slab has two functions: (1) it transfers load laterally to the joists, and (2) it serves as the top flange of the joists, which act as T-shaped beams that transmit the load to the beams running at right angles to the joists. The first floor of the building in Fig. 3−5 has a slab-and-beam design in which the slab spans between beams, which in turn apply loads to the columns. The column loads are applied to spread footings, which distribute the load over a sufficient area of soil to prevent overloading of the soil. Some soil conditions may require the use of pile foundations or other deep foundations. At the perimeter of the building, the floor loads are supported either directly on the walls as shown in Fig. 3−5, or on exterior columns as shown in Fig. 3−6. The walls are supported by a basement wall and wall footings. The choice of whether or not to have exterior

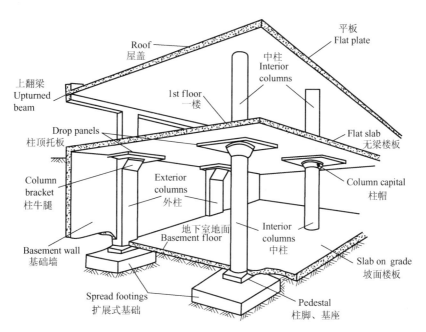

Fig. 3 – 6 Reinforced concrete building elements

columns is based on economic and architectural considerations. In most cases the construction scheduling is most efficient if the concrete frame is completely self-supporting and the masonry walls are built later.

The slabs in Fig. 3 – 5 are assumed to carry the loads in a north-south direction (see direction arrow) to the joists or beams, which carry the loads in an east-west direction to other beams, girders, columns, or walls. This is referred to as one-way slab action and is analogous to a wooden floor in a house, in which the floor decking transmits loads to perpendicular floor joists, which carry the loads to supporting beams, and so on.

The slab or plate type of structure shown in Fig. 3 – 6 is unique to reinforced concrete. Here the loads applied to the roof and the weight of the roof is transmitted in two directions to the columns by plate action. Such slabs are referred to as two-way slabs.

The first floor in Fig. 3 – 6 is a flat slab with thickened areas called drop panels at the columns. In addition, the tops of the columns are enlarged in the form of capitals or brackets. The thickening provides extra depth for moment and shear resistance adjacent to the columns. They also tend to reduce the slab deflections.

The roof shown in Fig. 3 – 6 is of uniform thickness throughout without drop panels or column capitals. Such a floor is a special type of flat slab referred to as a flat plate. Flat-plate floors are widely used in apartments because the underside of the slab is flat and hence can be used as the ceiling of the room below. Of equal importance, the forming for a flat plate is generally cheaper than that for flat slabs or for one-way slab-and-beam floors.

Precast concrete buildings generally follow one of two patterns: Frames composed of precast beams and columns with channel elements for floor members are widely used for offices, schools, and similar buildings; for apartment and hotel structures, one-way precast floor slabs supported by precast walls are frequently used.

参 考 译 文

第 3 课　建筑与建筑体系

1. 建筑

建筑学和房屋建筑并不一定是指同一事物。在房屋设计与建造中，应该了解如何将不同的材料、元素及构件组合在一起。但是，这种理解能够有助于人们建造房屋却不能给予质量保障。房屋建造的专业知识只是完成建造工作的关键因素之一。如果我们把建筑学称之为建筑艺术，那么除了实际的建造体系，我们还应该考虑下列概念体系：

(1) 建筑内部空间的定义、规模、比例及组织。

(2) 由不同比例与尺度体现的人类活动的顺序。

(3) 基于不同目的与用途的建筑空间的功能分区。

(4) 穿过建筑内部的水平和垂直通道。

(5) 建筑的感官质量：形状、空间、光线、颜色、质感及样式。

(6) 作为自然和建筑环境组成构件的建筑物。

本文中我们主要关注的是规定、组织并增强建筑的感官上及概念上排序的实际建造体系。

一个建筑体系可以被定义为由相互关联或相互依赖的部分集合而成，这些部分为共同目的而组成了更复杂的统一体。建筑可以被理解为是一些彼此相关且协调的系统以及子系统的实际体现，而且从整体上看，这些系统和子系统必须与建筑的空间结构和空间组织形式相关且协调（图 3-1）。

2. 建筑体系

1) 结构体系（图 3-2）

设计以及建造建筑结构体系的目的是支承并安全地将重力和水平荷载传递到地面且不超出构件的容许应力。

(1) 上部结构是指基础上部建筑的竖向延伸。

(2) 柱、梁及承重墙支撑地板和屋顶结构。

(3) 下部结构是指构成建筑基础的底部结构。

2) 围护体系

围护体系是指建筑物的外壳或围护结构，包括屋顶、外墙、窗户及门。

(1) 屋顶和外墙能够保护内部空间免受恶劣天气的影响，并控制水分、热量以及气流通过建筑的构造层次。

(2) 外墙和屋顶能够抑制噪声并为建筑物的居住者提供安全及隐私。

(3) 门能够提供实体通道。

(4) 窗户为光线、空气及视野提供路径。

(5) 内墙及隔墙将建筑物内部分成若干空间单元。

Lesson 3 Building and Building System

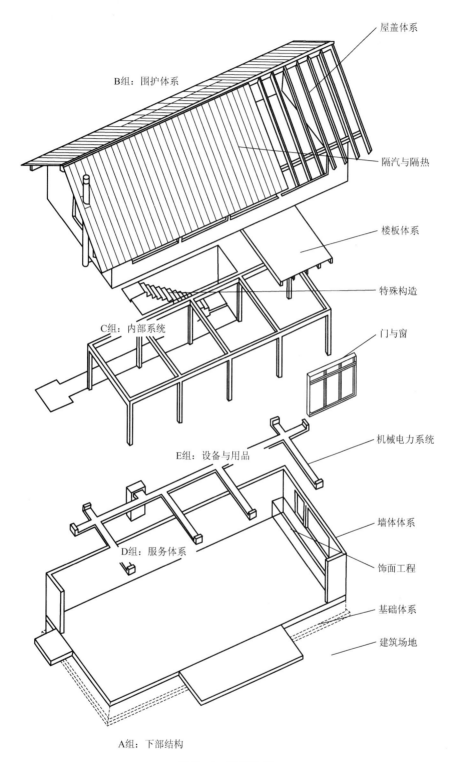

图 3-1 建筑体系

3) 机械体系

建筑物的机械体系能够为其提供基本服务。

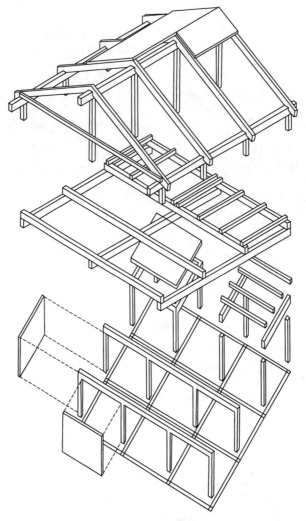

图 3-2 结构体系

（1）供水系统能够为人类的生活提供饮用水。

（2）污水处理系统能够将流体废物及有机物质清除出建筑物。

（3）供热、通风及空调系统能够调节建筑物内部空间的环境以为居住者提供舒适的环境。

（4）电力系统控制、计量及保护建筑物的电力供应，并将其安全地分配至电力、照明、安保及通信系统。

（5）中高与高层建筑物内的垂直运输系统将人与物由一层运送到另一层。

（6）消防系统能够监测并扑灭火灾。

（7）建筑物也需要污水处理和再循环利用系统。

3. 建造中建筑体系的考虑因素

在施工中选择、组装及整合不同建筑体系时，我们应该将下列因素纳入考虑范围：

1）性能要求

（1）结构相容性、整体化及安全性。

(2) 耐火性、火灾的预防及安全性。
(3) 结构构件的允许或者理想厚度。
(4) 通过建筑构件的热量及气流的控制。
(5) 水蒸气的迁移及冷凝的控制。
(6) 调整因沉降、结构挠度及温度和湿度变化导致的膨胀或收缩引起的建筑活动。
(7) 降噪、隔绝或隐蔽声音。
(8) 耐磨损、抗腐蚀及耐风化的能力。
(9) 饰面处理、整洁及维护要求。
(10) 使用安全。
2) 美学特质
(1) 建筑物与现场、临近物产及邻居之间预期的相互关系。
(2) 建筑外形、体量、颜色、图案、质地及细节的理想品质。
3) 法规限制
符合区域条例与建筑规范。
4) 考虑经济因素
(1) 初始成本包括材料、运输、设备及人工成本。
(2) 全生命周期成本不仅包括初始成本,而且包括维护和操作成本、能量损耗、有效寿命、拆除和更换成本及投资的利息。
5) 环境影响
(1) 通过选址和建筑设计节约能源和资源。
(2) 机械系统的能源功效。
(3) 高效资源以及无毒材料的使用。
6) 施工作业
(1) 安全要求。
(2) 容许公差及适当的调整。
(3) 符合行业标准和保证。
(4) 工厂与现场之间的作业划分。
(5) 劳动分工及建筑业的协调。
(6) 预算限制。
(7) 所需的施工设备。
(8) 所需的施工时间。
(9) 对恶劣天气的规定。

Lesson 4

Materials in Construction

1. Stages in the life of a building

The successful operation of materials in buildings requires an understanding of their characteristics as they affect the building at all stages of its lifetime as follows:

1) **Design**

Materials appropriate to the building design, function and environment must be selected; indeed, traditionally, consideration of the materials to be used was a coherent part of the design process. Today there is more choice but designers must be aware of the limitations as well as the opportunities associated with individual materials, and how they interact with each other.

2) **Construction**

Builders are legally obliged to "build well" and this includes a responsibility to use specified materials in the correct manner and to identify potential defects.

3) **Maintenance**

Although some materials are virtually maintenance-free, the majority require some form of care during their lifetime. Effective maintenance depends upon knowledge of how materials react with their environment over the planned lifetime. It may in the long term be much more expensive to maintain certain materials if their initial quality is poor.

4) **Repair**

A wide choice of repair systems is now available for many materials. Informed and efficient use of such systems hinges upon an understanding of decay processes, how to arrest decay, and how the repair materials interact with the original materials.

5) **Demolition/recycling**

These processes are increasingly being considered as a part of the overall selection procedure. The safety/environmental aspects of alternative materials are, in particular, now being included in assessments of overall suitability.

2. Materials performance

In each case, satisfactory operation of the building as a whole depends on the performance of the materials from which its components are made and on the interrelationships between them. ①Before assessing the suitability of any one material for a given situation, the performance requirements for that situation must be identified.

1) Stability (structural)

There are many aspects of structural performance; the main ones being as follows:

(1) Strength.

Strength may be defined as the ability to resist failure or excessive plastic deformation under stress.

There are several types of stress (Fig. 4 – 1). In general, the strength type selected should be as close as possible to that experienced in the structure, but even then, care is necessary with all types of strength measurement.

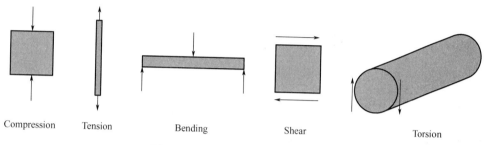

Fig. 4 – 1 Basic stress configurations

(2) Stiffness.

Stiffness is the ability of a material to resist elastic deformation under load.

Elastic deformation is deformation which is recovered when the load is removed. High deformations, even if elastic, may cause problems. It is normal practice, therefore, to check deflections as part of the structural design. Stiffness is normally measured by:

$$\text{elastic modulus}(E) = \frac{\text{applied stress}}{\text{strain caused by that stress}}$$

Since strain has no units (=fractional change of length), E values have the units of stress. Measured strains are usually small, typically less than 0.001, leading to large values of E. They are, therefore, usually measured in kN/mm^2.

(3) Toughness.

Toughness is the ability of a material to absorb energy by impact or sudden blow.

Strong materials are not always tough — for example, cast iron. Relatively weak materials can have high toughness — for example, leather. Fig. 4 – 2 illustrates a typical impact test such as that used to measure the behaviour of toughened glass.

(4) Hardness.

Hardness is resistance to indentation under stress.

Hardness is relevant to floor and wall surfaces (Fig. 4 – 3) and depends on a combination of strength and stiffness properties.

(5) Creep.

Creep is the effect of long-term stress, leading to additional distortion or failure.

Creep does not necessarily produce a safety risk though marked deflections in beams can appear alarming. In extreme cases, creep may lead to ultimate failure. Materials subject to creep are timber, clay, concrete, thermoplastics and, to a small extent, glass. Creep can be tested using rigs of the form shown in Fig. 4 - 4.

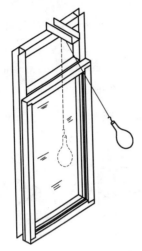

Fig. 4 - 2   Impact tester suitable for testing toughness of glass

Fig. 4 - 3   Hardness testing

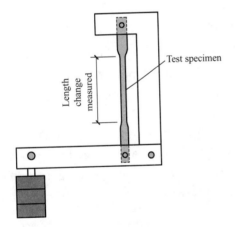

Fig. 4 - 4   Creep testing

(6) Fatigue.

Fatigue is the effect of load reversals such as vibrations which lead to failure at relatively low stresses.

All materials are subject to fatigue effects and fatigue may be the critical factor in design. The fatigue life of components is measured in cycles since each cycle adds to the level of damage. For metals, fatigue lives of several million cycles may be required in situations at risk.

2) Safety in fire

Evaluating fire hazard is a highly complex process and this is reflected in practice by the numerous BS tests for the different aspects of fire. The chief hazards can be summarized:

(1) Heat itself causes burns.

(2) Fire may endanger the structure.

(3) Many materials generate toxic fumes when they are heated.

(4) Materials often generate smoke, which makes breathing difficult and causes panic and disorientation as people try to escape.

3) Durability

A material may be durable in any one situation if it fulfils all its performance requirements, either for the planned lifetime of building, or for a shorter defined period where this is acceptable.

It is often very difficult to predict the durability of individual components. Also, since failures in a very small proportion of the items in use may be unacceptable, the only safe course of action may be to "over-design" them so that materials in the worst likely situation should be satisfactory. In consequence, many buildings last much longer than their design lifetime.

4) Safety in use and health

Public awareness of health and safety issues is increasing rapidly and the construction industry has come under scrutiny both from the point of view of safety of the construction team and the general public during the construction process and because the finished building might present a hazard to the occupants. Table 4-1 indicates some of the possible safety hazards posed by materials, together with recommendations of remedy.

Table 4-1 Common construction industry materials: risks

Substance	Situation	Risk	Remedy
Asbestos	Not currently used but may be present as insulation, especially in older buildings	Risk of lung cancer	Arrange for safe removal/disposal if found
Polyvinyl chloride (PVC)	Widely used for waste systems, windows, miscellaneous applications	Chief risk is in manufacture; small risk of toxic gas in fire	Stringent precautions during manufacture; normal fire precautions
Bitumen	Site application	Vapors of hot applied material may be hazardous	Take precautions during use

5) Environmental issues

Environmental matters are beginning to feature much more strongly in the assessment of building materials, the following key aspects of environmental performance will be considered:

(1) Embodied energy.

Energy should be regarded as a precious and finite resource and although it is relatively cheap at the present time there is an economic advantage of use of materials which take less energy to convert them from raw materials into the final, in-situ product. In seeking new materials, emphasis should be given to those materials which consume less energy in their manufacturing process, and such materials should become increasingly competitive when energy costs rise. The process can be quite complex since costs must be assessed at each stage.

(2) Recycling potential.

This may conveniently be described as a hierarchy of Rs, as shown in Table 4-2. Only when none of these is feasible should disposal be considered.② Materials/situations in which disposal is the only practical option should be regarded as failures in an environmental sense.

Recycling may have various goals:

① Reduction of raw materials consumption.

② Reduction of fuel consumption in manufacture.

③ Reduction in waste generation.

Table 4-2 Recycling options

Option	Comments
Reduce	Problems can be avoided at source if wastage in production is reduced and materials have inherently low maintenance characteristics
Reuse	This implies minimum recovery/reprocessing cost—for example, roofing tiles and brick bedded in a lime mortar
Recycle	Some manufacturing is involved—for example, lead, steel, glass. Solid wood could be reprocessed into chipboard
Recover energy	This may still be worthwhile if combustible materials such as timber, plastics and rubber are incinerated and the heat recovered, say, for district heating schemes
Disposal	Should be the last resort

(3) Environmental management.

There are calls for sustainable construction, which is a subset of a sustainable society. Such a society should be able to continue to operate without compromising prospects for operation of future societies. In environmental terms, this means avoiding the following:

① Depletion of finite resources.

② Adversely affecting the environment by pollution or waste.

③ Adversely affecting the environment by energy emission.

④ Adversely affecting the environment in a broader sense, such as upsetting the ecobalance of wildlife.

Ⅰ. New Words

1. coherent *adj.* 连贯的，一致的，清晰的，粘在一起的

2. identify *vt.* 确定，鉴定，识别，辨认

3. virtually *adv.* 事实上，几乎，实质上

4. hinge *n.* 铰链，折叶，同关键，中枢

5. decay *n.* 衰退，衰减，腐烂；*vt.* 使衰退，使衰落，使腐败，

6. demolition *n.* 拆除（等于 demolishment），破坏，毁坏

7. alternative *adj.* 供选择的，交替的；*n.* 二中择一，供替代的选择

8. stiffness *n.* 劲性，刚度，坚硬，顽固

9. deformation *n.* 变形
10. deflection *n.* 挠度，挠曲，偏差
11. modulus *n.* 模量，模数
12. fractional *adj.* 部分的，分数的，小数的
13. toughness *n.* 韧性，韧度
14. toughen *vt.* 使变坚韧，使变坚强
15. indentation *n.* 压痕，刻痕，呈锯齿状
16. combination *n.* 结合，组合
17. creep *n.* 徐变，蠕变
18. distortion *n.* 变形，扭曲，曲解
19. thermoplastic *n.* 热塑性塑料；*adj.* 热塑（性）的
20. fatigue *n.* 疲劳，疲倦，疲乏
21. reversal *n.* 逆转，反转
22. disorientation *n.* 迷失，迷惑
23. predict *vt.* 预报，预告，预言
24. remedy *vt.* 补救，治疗，纠正；*n.* 补救，治疗，赔偿
25. hierarchy *n.* 层级，等级制度
26. subset *n.* 子集
27. sustainable *adj.* 可持续的，可以忍受的
28. compromise *vt.* 妥协，让步
29. depletion *n.* 消耗，损耗

Ⅱ. Phrases and Expressions

1. structural performance   结构性能
2. plastic deformation   塑性变形
3. elastic modulus   弹性模量

Ⅲ. Notes

① "In each case, ... from which ..." 是介词 from 引导的定语从句，修饰限定 "the materials, ... and on the interrelationships between them."，与 "depends on the performance of the materials..." 并列。

② "Only when none of these is feasible should disposal be considered." 此句为倒装句，从句中因将 "Only" 提前，所以主句 "disposal should be considered" 将 "should" 提前形成了倒装句。

Ⅳ. Exercises

1. Explain the difference between the terms "strength" and "stiffness". Give examples of materials which have the following performace.

(a) high strength and high stiffness　(b) low strength and high stiffness

(c) high strength and low stiffness　(d) low strength and low stiffness

2. Write down the equation to measure stiffness.

3. Recycling potential may conveniently be described as a hierarchy of Rs. 4Rs represents _____, _____, _____, and _____.

Ⅴ. Expanding

Know about the following terms related to the materials and the properties.

1. organic material　有机材料
2. inorganic materials　无机材料
3. composite material　复合材料
4. waterproof material　防水材料
5. thermal insulation material　隔热材料
6. sound absorption material　吸声材料
7. mechanical property　力学性质
8. physical property　物理性质
9. apparent density　表观密度
10. stacking density　堆积密度
11. compactness　密实度
12. porosity　孔隙率
13. filling rate　填充率
14. voidage　空隙率
15. hydrophilic　亲水性
16. hydrophobicity　憎水性
17. water-absorbing quality　吸水性
18. hygroscopicity　吸湿性
19. waterproof　耐水性
20. anti-permeability　抗渗性

Ⅵ. Reading Material

More Recent Materials

A consideration of how materials could be improved follows naturally from Table 4-3. Within the metals group, it would be fair comment that high-performance metals already exist—for example, steel wire is available with strengths approaching $2000N/mm^2$. The main problem with metals is: high manufacturing energy requirement, high density, high cost.

Metals undoubtedly fulfil a vital function in construction, but in view of the above, use should be restricted to situations where requirements cannot be met by other materials.

Table 4-3 Common construction materials: factors limiting strength

Material	Bond type	Factor limiting the strength
Metals	Metallic crystals	Dislocations; grain boundaries
Cements	Microscopic ionic/covalent interlocking fine crystals	Non crystalline areas; creaks; pores (voids)
Thermosets	3D amorphous ionic/covalent molecules	Limited numbers of bonds
Thermoplastics	Polymer chains with interlocking/van der Waals forces	Van der Waals forces

In terms of other materials groups, weaknesses could be broadly classified as:

(1) Poor tensile strength and impact properties of materials such as concrete.

(2) Low stiffness of plastics.

Progress has been made in addressing these deficiencies. Two techniques will be briefly described.

1. High-performance concretes

Attempts to improve the tensile strength of concretes must focus upon the main cause of weakness. Cement hydrate contains pores and most concretes also contain pores resulting from the evaporation of excess water after hardening.

Efforts have been made to control both the total porosity of the material and the size of pores.

2. Control of porosity

It is well known that reducing the water/cement ratio of concrete increases strength since:

(1) Unhydrated cement is stronger than hydrated material.

(2) Less water in the mix results in fewer pores in the long term.

Very high strength can be obtained by use of very low water/cement ratios. These concretes are, however, very difficult to handle in the fresh state. Additives such as super plasticizers and microsilica can permit concrete compaction to be achieved, albeit with difficulty. Concretes with compressive strengths of over $100N/mm^2$ and tensile strengths of $15N/mm^2$ can be obtained in this way, but such products are best suited to factory production. They would permit a reduction in section sizes but would still need to be reinforced in situations of high-tensile stress or impact.

3. Reduction of size of pores: MDF cement

Experiments have been carried out to find ways of reducing the size of pores since it is known that larger pores are mainly responsible for strength reduction.

The origin of the larger pores or defects appears to be packing problems in cement particles caused by friction between them. It is now possible to eliminate these in certain special Portland and high-alumina cements by the addition of water-soluble polymers, such as polyvinyl alcohol/acetate, which reduce particle friction, allowing defects to be removed by rolling low water/cement ratio mixes into a dough-like material until macro-defects are no larger than 0.1mm. Ordinary plasticizers and super-plasticizers cannot be used for this process because they do not appear to reduce particle friction in such mixes.

Residual porosity can be removed by heating the moulded material to 80℃ for a short time. This causes a shrinkage of almost 10 percent in the material as it hydrates, corresponding to a large reduction in porosity. The final product has a flexural strength of around $150N/mm^2$ with an elastic modulus of about $50kN/mm^2$, indicating a strain capability of around 3000×10^{-6}; rather larger than the yield strain of mild steel.

Fig. 4-5 shows examples of small artefacts produced from MDF cement. Widespread use of such products has yet to occur, partly because properties deteriorate on wetting.

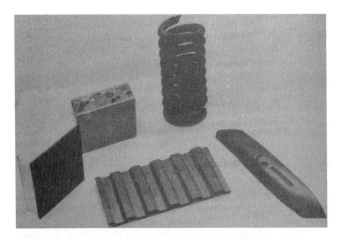

Fig. 4-5 Some simple artefacts made from MDF cement

4. Fibrous composites

Fibres may address the shortcomings in cement-based materials and in plastic as they have the distinct structural advantage of being of small diameter. It has been explained above that practical strengths are often limited by defects in materials. It is found that fibres have, in general, many fewer defects than bulk equivalents, due at least in part to the fact that stresses in production (for example, thermal stresses) are greatly reduced and also that there is less scope for large defects in samples of small size. The tensile strength of glass fibre is increased, for example, from about $100N/mm^2$ to around $2000N/mm^2$ by reducing the size of individual filaments to about $10\mu m$. (It should be noted that such glass fibres are much smaller than those designed for thermal insulation.)

The requirements for improvement of the properties of cement-based materials and plastics are different. In order to improve the tensile performance of concretes, fibres will need to:

(1) grip the matrix well;
(2) have high E value in order to carry as much stress as possible;
(3) have sufficient strength to carry imposed stresses.

Fig. 4 – 6 shows how fibres need to carry loads across a crack. It is found that relatively small volumes of high-performance fibres can achieve these ends and in any case it is not usually easy to incorporate large quantities of fibres into cement-type materials.

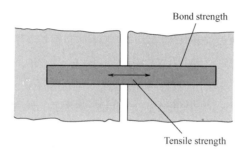

Fig. 4 – 6 Use of fibres to bond cracks together

Fig. 4 – 7 shows a special steel fibre-reinforced mortar being used to transfer load between reinforcing bars. The small beam shown contained 12-mm bars lapped by only 80mm and carried a load as high as that carried by a similar continuously reinforced beam. Such arrangements permit efficient structural coupling of precast units.

Fig. 4 – 7 Use of steel fibre reinforced concrete to achieve stress transfer between lapped reinforcing bars

As far as plastics are concerned, fibres are usually incorporated to stiffen and strengthen them. This requires in essence that fibres:

(1) Have high E values relative to the plastic.
(2) Be used in the highest volume fractions possible.

The use of a few percent of fibres, as was the case when reinforcing cement-type materials, would be unlikely to be very effective. The following types are illustrations of fibres fulfilling these requirements:

Glass　　　　　　　　　　　　E value, $70kN/mm^2$
Kevlar (polyamide)　　　　　　E value, $100kN/mm^2$
Carbon　　　　　　　　　　　E value, $200kN/mm^2$

Of these, glass fibres are in most common use for reasons of cost.

参 考 译 文

第4课　建 筑 材 料

1. 建筑物寿命期内的各个阶段

建筑材料的成功运用需要了解其特性，因为材料会在以下方面对建筑物寿命期的各个阶段产生影响。

1）设计

必须选择适合建筑设计、功能和环境的材料；事实上，传统来说，对所使用材料的考虑是设计过程的重要组成部分。今天有更多的选择，但设计师必须注意某些限制，以及与个别材料有关的因素，以及它们之间是如何相互影响的。

2）施工

建设者在法律上有义务"建设优质建筑"，这包括有责任和义务以正确的方式使用规定的材料并辨别其潜在的缺陷。

3）维护

尽管一些材料实际上不需要维护，但大多数材料在其使用寿命期间需要进行某种方式的保养。进行有效的维护取决于材料在设计使用寿命期内与环境相互作用方面的知识。从长远来看，如果某些材料的初始质量较差，其维护费用可能会更高。

4）修复

现在用于多种材料的修复系统可以有更多的选择。对这种系统的了解和有效利用取决于对退化过程的了解，如何阻止退化，以及修复材料与初始材料间的相互作用如何。

5）拆除/回收

这些过程越来越多地被视为整个材料选择过程的一部分。特别是所选择的材料在安全/环境方面的因素现在被包括在总体适用性的评估中。

2. 材料性能

在每种情况下，建筑物整体达到令人满意的作业取决于制成其构件的材料性能和材料间的相互关系。在评估任何一种材料对给定情况的适用性之前，必须确定该情况对材料性能的要求。

1）稳定性（结构的）

结构性能有很多方面，主要内容如下。

（1）强度。

强度可以定义为在应力作用下抵抗失效或过量塑性变形的能力。

有几种应力类型如图4-1所示。通常，所选择的强度类型应尽可能接近结构中所遭受的强度类型，但即使如此，对所有类型的强度测定也需要小心。

Lesson 4　Materials in Construction

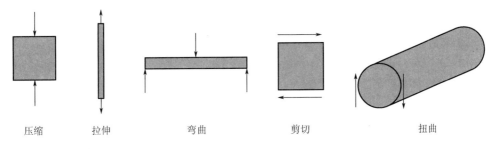

压缩　　　　拉伸　　　　弯曲　　　　剪切　　　　　　扭曲

图 4-1　基本应力形式

（2）刚度。

刚度是载荷作用下材料抵抗弹性变形的能力。

弹性变形是当去除载荷时可恢复的变形。较大的变形，即使是弹性的，也可能引起一些问题。因此，作为结构设计的一部分，测定挠度是常规做法。刚度通常通过以下公式来测定：

$$弹性模量(E)=\frac{产生的应力}{由应力产生的应变}$$

因为应变没有单位（＝单位长度的变形），E 值具有应力的单位。测量到的应变通常很小，一般小于 0.001，导致 E 值很大。因此，它们通常以 kN/mm^2 为单位进行测量。

（3）韧性。

韧性是指材料通过冲击或突然锤击来吸收能量的能力。

坚固的材料并不总是坚硬的，例如铸铁。相对较弱的材料也可以具有高韧性，例如皮革。如图 4-2 所示为典型的冲击试验，例如用于测量钢化玻璃性能的试验。

（4）硬度。

硬度是在应力下对压痕的抵抗力。

硬度与楼板或墙体表面有关（图 4-3），取决于强度和刚度特性的组合。

（5）蠕变。

蠕变是由长期应力的影响造成的，会导致附加变形或失效。

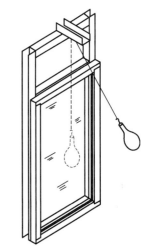

图 4-2　适用于测试玻璃韧性的冲击试验机

尽管梁中产生明显的挠度可能会出现警报，但是蠕变不一定会产生安全风险。在极端情况下，蠕变可能导致最终失效。易蠕变的材料包括木材、黏土、混凝土、热塑性塑料，以及在小范围内蠕变的玻璃。蠕变可用图 4-4 中所示形式的装置进行测试。

（6）疲劳。

疲劳是由负载逆转（例如相对较低应力下可造成失效的振动）的影响造成的。

所有材料都会受到疲劳影响，疲劳可能是设计的关键因素。构件的疲劳寿命要循环测量，因为每次循环都会增加损坏等级。对于金属，在有风险的情况下可能需要有承受几百万次循环的疲劳寿命。

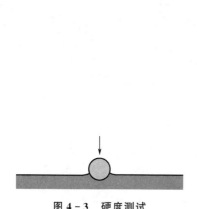

图 4-3 硬度测试

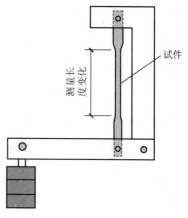

图 4-4 蠕变试验

2) 火灾安全

评估火灾危险是一个非常复杂的过程,在实践中要通过许多 BS 测试来反映火灾的不同方面。主要危害可概括为:

(1) 热本身导致的灼伤。

(2) 火可能危及结构。

(3) 许多材料在加热时产生有毒的烟雾。

(4) 材料经常产生烟雾,这使得人们在试图逃跑时呼吸困难并且引起恐慌和迷失方向。

3) 耐久性

如果材料满足其所有性能要求,或者满足建筑物的设计寿命要求,或者满足较短的可接受的限定时间要求,那么该种材料在任何一种情况下都被认为是耐久的。

预测各个构件的耐久性通常是非常困难的。此外,构件发生非常小范围的失效也是不可接受的,采取唯一安全的行动就是"有裕量地设计",它们使得材料在可能最坏的情况下都能满足要求。因此,许多建筑物的使用寿命比其设计使用寿命长很多。

4) 使用安全和健康安全

公众对健康和安全问题的意识正在迅速提高,并且从建筑团队和普通大众的安全角度考虑建筑业在建造过程中要受到审查,因为已建建筑可能对居住者造成危害。表 4-1 列出了材料可能产生的一些安全隐患,以及补救的建议。

表 4-1 建筑业常用材料:风险

物 质	状 况	风 险	补 救
石棉	目前未使用,但可能用于绝缘,特别是在老建筑物中	肺癌风险	发现后要基于安全合理安排移除/废弃处理
聚氯乙烯(PVC)	广泛用于废物处理系统,窗或其他等方面的应用	主要风险在于制造。在火灾中有产生有毒气体的小风险	在制造时要有严格的预防措施,正常的防火措施
沥青	现场应用	材料热处理的蒸汽可能是有害的	使用时应采取防范措施

5）环境问题

环境问题在开始对建筑材料评估中就超显著作用，环境性能要考虑以下关键方面：

（1）实体能源。

能源应该被视为一种宝贵的和有限的资源，虽然目前相对便宜，材料的应用具有经济优势，这种材料只消耗较少的能量就能将其从原料转化为最终的现场产品。在寻求新材料时，应重视那些在其制造过程中消耗较少能量的材料，并且这些材料在能量成本上升时应当变得更具竞争力。这个过程可能相当复杂，因为在每个阶段都必须要评估成本。

（2）回收潜力。

这可以方便地描述为 Rs 的层次结构，如表 4-2 所示。只有当这些都不可行时，才应考虑废弃处理。对于材料/环境而言，废弃处理被视为环境意义上的失效时的唯一现实选择。

回收可能有多种目的：

① 减少原材料的消耗。

② 减少制造中的燃料消耗。

③ 减少废物的产生。

表 4-2 资源回收选项

选 项	注 释
减少	如果减少生产中的浪费并且材料本身具有低成本维护的特性，则可以在源头避免问题
重复利用	这意味着最低的回收/再加工成本，例如，固定在石灰砂浆中的屋顶瓦和砖
回收	涉及一些制造业，如铅、钢、玻璃。实木可以重新加工成刨花板
回收能源	如果将可燃材料（如木材、塑料和橡胶）焚烧并且热量回收，如用于区域供热方案，这些做法还是值得的
废弃处理	应该是最后的手段

（3）环境管理。

要求可持续建设，是可持续发展社会的一个方面。这样的社会应该能够继续运转，而不会损害未来社会的经营前景。在环境方面，这意味着避免以下情况：

① 有限资源的消耗。

② 污染或浪费对环境产生不利影响。

③ 能源排放对环境产生不利影响。

④ 在更广泛的意义上对环境产生不利影响，如破坏野生动物的生态平衡。

Lesson 5

Concrete

1. Introduction

Concrete is essentially a mixture of cement, aggregates and water. Other materials added at the mixer are referred to as "admixtures". Materials based on cement have the following general attractions:

(1) Low cost.
(2) Flexibility of application—for example, mortars, concretes, grouts, etc.
(3) Variety of finishes obtainable.
(4) Good compressive strength.
(5) Protection of embedded steel.

The following are some disadvantages/problems:

(1) Low tensile strength/brittleness.
(2) Rather high density (though lower density types are available).
(3) Susceptibility to frost/chemical deterioration (depending on type).

2. Cements

The most important are Portland cements; so named because of a similarity in appearance of concrete made with these cements to Portland stone.

Portland cements are hydraulic, that is, they set and harden by the action of water only.

1) Manufacture of Portland cements

Portland cements are made by heating a finely divided mixture of clay or shale and chalk or limestone in a kiln to a high temperature — around 1500℃, such that chemical combination occurs between them. About 5 percent gypsum (calcium sulphate) is added to the resulting clinker in order to prevent "flash" setting and the final stage involves grinding to a fine powder.

2) Composition of Portland cements

Although the familiar grey powder may appear to have a high degree of uniformity, it is important to appreciate that Portland cement is a complex combination of the minerals contained originally in the clay or shale and the calcium carbonate which constitutes limestone or chalk. The chief compounds are listed below, together with commonly used abbreviations.

(1) Clay or shale.

SiO_2 silica (silicon oxide), abbreviated S.

Fe_2O_3 ferrite (iron oxide), abbreviated F.

Al_2O_3 alumina (aluminium oxide), abbreviated A.

(2) Limestone or chalk.

$CaCO_3$ (calcium carbonate on heating gives CaO quicklime), abbreviated C.

<u>Chemical analysis of cement clinker shows that four chief compounds are present, dicalcium silicate (C_2S), tricalcium silicate (C_3S), tricalcium aluminate (C_3A), and tetracalcium aluminoferrite (C_4AF).</u>[1]

The properties of Portland cements vary markedly with the proportions of these four compounds, reflecting lots of differences between their individual behavior. The proportion of clay to chalk (usually approximately 1 : 4 by weight) must be very carefully controlled during manufacture of the cement, since quite small variations in the ratio produce relatively large variations in the ratio of dicalcium silicate to tricalcium silicate. A greater proportion of chalk favors the formation of more of the latter, since it is richer in lime. This would lead to a cement with more rapid early strength development. The tricalcium aluminate and the tetracalcium aluminoferrite contribute little to the long-term strength or durability of Portland cements but would be difficult to remove and, in any case, are useful in the manufacturing process. They act as fluxes, assisting in the formation of the silicate compounds, which would otherwise require higher kiln temperatures.

The properties of cement are dependent not only on its composition but also on its fineness. This is because cement grains have very low solubility in water. This could be demonstrated by mixing a small amount of cement with water in a beaker and leaving for some hours, stirring occasionally.

Ordinary Portland cement (OPC) is the cement best suited to general concreting purposes, it is the lowest priced cement and combines a reasonable rate of hardening with moderate heat output. Other types of cement are, however, available, such as rapid-hardening Portland cement, sulphate-resisting Portland cement, low-heat Portland cement, Portland blast-furnace cement, and pulverized fuel ash (PFA) in cements.

3. Aggregates for concrete

Aggregates are used in concrete for the following reasons:

(1) They greatly reduce cost.

(2) They reduce the heat output per unit volume of concrete and therefore reduce thermal stress.

(3) They reduce the shrinkage of the concrete.

(4) They help to produce a concrete with satisfactory plastic properties.

There may be other reasons for which "special" aggregates would be employed, for example:

(1) Low-density concrete ($2000kg/m^3$ or less) to decrease foundation loads, increase thermal insulation and reduce thermal inertia.

(2) High-density concrete ($2600kg/m^3$ or more) as required (for radiation shielding).

(3) Abrasion-resistant concretes for floors (granite aggregates).

(4) Improved fire resistance (limestone, lightweight aggregates).

(5) Decorative aggregates — for example, crushed granite is available in several different colors which can be revealed by use of an exposed aggregate finish.

4. Water for concrete

Water is the chemical means by which cement is converted from a powder into a hardened material with strength and durability. It therefore needs to be of appropriate quality and two main aspects may need to be considered.

1) Organic contamination

Water which is in contact with organic matter such as vegetation can be contaminated by organic acids. These can reduce the rate of hydration by increasing the pH value (acidity) of the wet concrete. Water which appears green due to the presence of algae can lead to air entrainment which reduces strength.

2) Dissolved salts

Sea water, for example, may contain dissolved salts such as sulphates which can react with the hydrated cement, or chlorides which tend to accelerate hydration and increase the risk of corrosion of embedded steel. Where chlorides are present the salt content of the water should be added to that of the aggregates to obtain a combined total figure. This should then be checked against allowable values.

In general, clear, flowing water or drinking water should be suitable for production of concrete but, if in doubt, cubes can be made to the required specification and checked against similar cubes made with distilled water.

5. Principles of proportioning concrete mixes

Consideration has already been given to properties of cements and aggregates and it is necessary to identify the important properties of concrete and to examine the effect of materials types and proportions on them.

Perhaps the most important single term in understanding how mix parameters affect concrete properties is the water/cement ratio.

$$\text{water/cement ratio} = \frac{\text{mass of water in a concrete sample}}{\text{mass of cement in the sample}}$$

Note that the free-water/cement ratio is the most important ratio since it is this that affects strength and durability.[②] Free-water/cement ratio would be based on the free-water content of the mix—that is, water absorbed in the aggregates is disregarded.

A further term traditionally used is the aggregate/cement ratio[③].

$$\text{aggregate/cement ratio} = \frac{\text{mass of aggregate in a concrete sample}}{\text{mass of cement in that sample}}$$

For most concrete mixes the aggregate/cement ratio would be in the range 4—10, small ratios indicating rich, expensive mixes and high ratios lean, cheaper mixes. Although mortars are still commonly specified in terms of aggregate (sand)/cement ratios the richness of concrete mixes is now more commonly specified in terms of cement content per cubic metre. For example, an aggregate/cement ratio of 4 would correspond to about 450kg of cement per cubic metre while an aggregate/cement ratio of 10 would correspond to about 200kg of cement per cubic metre. One advantage of expressing the richness of mixes in terms of cement content is that the cost of the concretes of varying richness can be relative easily compared, cement being the most costly item.

Ⅰ. New Words

1. cement *n.* 水泥，接合剂
2. aggregate *n.* 合计，集合体，骨料；*adj.* 集合的，合计的；*vt.* 集合，合计
3. admixture *n.* 混合，添加物，掺和剂
4. attraction *n.* 吸引，吸引力
5. mortar *n.* 砂浆，灰浆
6. grout *n.* 水泥浆，石灰浆，薄浆
7. brittleness *n.* 脆性，脆度
8. density *n.* 密度
9. susceptibility *n.* 敏感性
10. deterioration *n.* 恶化，退化
11. clay *n.* 黏土，泥土
12. shale *n.* 页岩，泥板岩
13. chalk *n.* 白垩岩；粉笔
14. limestone *n.* 石灰岩，石灰石
15. kiln *n.* (砖，石灰等的) 窑，炉
16. gypsum *n.* 石膏
17. calcium *n.* 钙
18. sulphate *n.* 硫酸盐
19. clinker *n.* 渣块，炼砖
20. grind *vt.* 磨成 (粉末)，磨碎
21. powder *n.* 粉，粉末，尘土
22. carbonate *n.* 碳酸盐
23. silica *n.* 二氧化硅
24. silicon *n.* 硅
25. oxide *n.* 氧化物
26. ferrite *n.* 铁酸盐，铁氧体
27. alumina *n.* 氧化铝，矾土
28. quicklime *n.* 生石灰

29. dicalcium *n.* 二钙化物

30. silicate *n.* 硅酸盐

31. tricalcium *n.* 三钙

32. aluminate *n.* 铝酸盐

33. tetracalcium *n.* 四钙

34. flux *n.* 流量，流出；*vt.* 使熔化，流出

35. solubility *n.* 溶解度，可溶性

36. stir *vt.* 搅拌；*vi.* 搅动

37. pulverize *vt.* 粉碎，使成粉末，研磨

38. shrinkage *n.* 收缩，减低

39. shield *n.* 庇护，包庇，掩盖

40. abrasion *n.* 磨损，磨耗，擦伤

41. granite *n.* 花岗岩

42. contaminate *vt.* 污染，弄脏，沾染

43. acid *n.* 酸；*adj.* 酸的

44. algae *n.* 藻类，水藻，海藻

45. dissolve *vt.* 使溶解，使分解，使液化

46. chloride *n.* 氯化物

47. corrosion *n.* 腐蚀，侵蚀，腐蚀作用的生成物（如锈）

48. distill *vt.* 蒸馏，提取

Ⅱ. Phrases and Expressions

1. compressive strength　抗压强度
2. tensile strength　抗拉强度
3. Portland cement　波特兰水泥
4. hydration reaction/hydration　水化反应
5. water/cement ratio　水灰比
6. aggregate/cement ratio　集料与水泥比（质量比）

Ⅲ. Notes

① 句中的"dicalcium, tricalcium, tetracalcium"是合成词，前缀 di-，tri-，tetra- 分别来自希腊语"二、三、四"之意，与"calcium"合成新词。合成词多是专业英语的一大特点，合成词用希腊词根和拉丁词根的比例较大。专业英语的另一大特点是缩写使用频繁，如 C_2S, C_3S, C_3A, C_4AF。

② "Note that… since it is this that affects strength and durability." 中主句"Note that…"使用了祈使句，表示请求、建议，因语气较弱，用句号结尾，专业英语中使用祈使语句很常见。另外，since 引导的原因状语从句，使用了强调句型，即"it is＋被强调部分＋that＋…"，强调的是主语"this"，其中"this"指代"the free-water/cement ratio"。

③ 文中"the aggregate/cement ratio"为集料与水泥的质量比，不要与混凝土拌制时"浆骨比"概念混淆，浆骨比是水泥浆与骨料用量之间的对比关系，即（水＋水泥用量）：（砂子＋石子用量）。

Ⅳ. Exercises

Fill in the blanks with the information given in the text.

1. Concrete is essentially a mixture of _____, _____ and _____.
2. Portland cements are _____, that is, they set and harden by the action of water only.
3. Chemical analysis of cement clinker shows that four chief compounds are present, _____, _____, _____, and _____.
4. The equation of the water/cement ratio is _____.
5. The equation of the aggregate/cement ratio is _____.

Ⅴ. Expanding

Remember the following terms related to the construction materials.

1. heavyweight concrete; heavy concrete   重混凝土
2. normal concrete   普通混凝土
3. light concrete   轻混凝土
4. cement concrete   水泥混凝土
5. asphalt concrete   沥青混凝土
6. gypsum concrete   石膏混凝土
7. polymer concrete   聚合物混凝土
8. waterproof concrete   防水混凝土
9. heat-resistant concrete   耐热混凝土
10. acid-resistant of concrete   耐酸混凝土
11. radiation shield concrete; anti-radiation concrete   防辐射混凝土
12. expansive concrete   膨胀混凝土
13. aerated concrete; aeroconcrete   加气混凝土
14. fly ash concrete   粉煤灰混凝土
15. fiber concrete; fiber reinforced concrete   纤维混凝土
16. silica fume concrete   硅灰混凝土
17. slag concrete   矿渣混凝土
18. silicate concrete   硅酸盐混凝土

Ⅵ. Reading Material

Durability of Concrete

In the 1980s, there was increasing concern over the rate of deterioration of some concrete structures. This, and the fact that concrete buildings can be very expensive to repair, together with a revitalized steel industry, led to a decline in the use of concrete for some applications. In most instances, deterioration of concrete could be linked to failure to implement existing technology due, for example, to a lack of supervision, though, in a few instances, decay has been the result of inadequate understanding of the behaviour of the material. There is now a greater awareness of potential problems and greater confidence in the performance of concrete generally.

Almost all forms of deterioration are the result of water ingress. The need to keep water out of structures, or to take special care where this is not feasible, cannot be overstated. The action of water may be two-fold:

(1) Physical attack, where water itself crystallizes on freezing or where drying out causes damage due to crystallization of dissolved salts in the water.

(2) Chemical damage. Most forms of chemical damage are associated with water since water is essential for the formation of ions—the chemically active form of salts or other materials.

The main types of deterioration are given below.

1. Frost attack

There are three prerequisites for frost attack:

(1) A permeable form of concrete able to admit water.

(2) The presence of water (though materials need not be saturated to be affected by frost).

(3) Temperatures below 0℃.

The mechanism of attack has long been attributed to the 10 percent expansion of absorbed water on freezing, though it is now known that further damage can result from movement of water within concrete on cooling below 0℃. Ice builds up in large pores and cracks and may cause very large expansion in such positions, while other areas become dryer (desiccation). Perhaps the extreme example of this is frost heave, as found in soils where large ice lenses can be obtained, Frost damage in concrete exacerbated as follows:

(1) Horizontal surfaces tend to absorb more water in wet conditions, take a longer time to dry and also cool quicker by radiation to the sky on a clear night.

(2) Very low temperatures increase the extent of migration of water and result in freezing to greater depths in the concrete.

(3) Repeated freezing and thawing (rather than prolonged steady freezing) add to the damage. Once freezing has occurred no further damage is done at a steady low temperature.

However, a thaw followed by a further frost will start another cycle of damage. Hence climates such as those found in the northern UK could be more damaging than colder climates where thawing is less frequent and, in any case, precipitation in such climates is more likely to be in the form of snow which, being solid, cannot penetrate and therefore cannot damage concrete until thawing occurs.

(4) Weak, permeable concretes which absorb water more readily are much more vulnerable to attack.

(5) De-icing salts melt the ice but add to saturation of the concrete. They then crystallize on drying, adding to the damage.

Most of the measures required to avoid frost attack will be evident from the above, though it should be added that use of air-entraining agents has been found to be particularly effective in avoiding damage to surfaces at risk. It is important to appreciate that new concrete with its incomplete hydrate structure and high permeability is particularly at risk from frost, and must be protected by an insulating material until it has sufficient maturity to resist frost.

2. Sulphate attack

Sulphates are often found in the ground, mainly associated with clay soils. These have low permeabilities, which means there is little chance of sulphates being washed out. To be active the sulphates must be in solution so that the risk of attack depends not only on the salt content of a soil but on the presence and movement of moisture. The principal mechanism is one in which sulphate ions react with hydrated C_3A in the cement, producing an expansive product, calcium sulphoaluminate, which disrupts the concrete. Sulphates commonly occur in three forms — calcium, sodium and magnesium — which pose quite different risks. Calcium has low solubility and does not constitute a high risk. Of the other two, magnesium is more harmful because:

(1) The reaction product is insoluble, precipitating out of solution. This avoids "chemical congestion" in the solution, which would tend to reduce the rate of attack.

(2) Magnesium sulphate also reacts with the C_3S hydrate in cement.

When estimating the risk of sulphates it is therefore important to identify the type of sulphates present. Total sulphate content could be misleading if it were mainly in the form of calcium sulphate. Careful sampling is also necessary since levels may vary widely from place to place and ground water may be diluted during sampling or after wet weather. The best procedure is normally analysis of ground water or water obtained from a 2 : 1 water-soil mixture.

For ground water containing up to 1.4 g/litre of SO_4 ions, OPC can be used provided a cement content of at least 330kg per cubic metre is used with a water cement ratio not higher than 0.5.

At sulphate levels up to 6.0mg/litre without high magnesium concentrations, use of pozzolanas at high replacement levels, or sulphate-resisting Portland cement, would be necessary.

At higher sulphate levels, or when high magnesium levels are found, the use of sulphate-resisting Portland cement in conjunction with a waterproof coating would be essential.

In all cases it is of paramount importance that attention is given to achieving full compaction of a good-quality impermeable concrete.

3. Crystallization damage

This may occur in concretes which are resistant to chemical attack. Sulphates (or other salts) may be admitted by permeable concretes and if, at a later stage, drying occurs, these salts crystallize, causing damage similar to that caused by frost. The condition may be severe when concrete is subject to wetting and drying cycles — as, for example, in sea walls around the high-tide mark. Attack may also be serious if one area of a structure is saturated while an adjacent area is dry — salts migrate towards the dry area, causing extensive crystallization in the region of drying. This type of damage may also be avoided by the use of low water/cement ratio concretes, fully compacted to produce low permeability.

4. Alkali-silica reaction (ASR)

Although relatively uncommon at the moment ASR, sometimes called "concrete cancer", has evoked a strong public response because it is a relatively new form of deterioration and there is no effective repair technique for structures that are seriously affected. ASR occurs when three conditions are met:

(1) The concrete has a very high pH value (over 12) resulting from the presence, in quite small quantities, of sodium and potassium alkalis originating from the oxides of these elements if present in the cement (high-alkali cements).

(2) Aggregates contain reactive forms of silica such as found in opal, tridymite, some flints and cherts.

(3) The presence of water.

The reactive components of the aggregates form a gel, leading to expansion and clisruption of the concrete, often producing "map" cracking. The reaction is slow and the cracking usually takes 10 — 20 years to become evident. There is a "pessimum" level of reactive aggregate of about 2 percent since smaller quantities do not form sufficient gel while larger quantities "swamp" the excess alkali present. Once formed, cracks admit further water, accelerating the rate of deterioration. ASR can be avoided by applying one or more of the following:

(1) Use of low-alkali cements. For example, low alkali versions of OPC containing less than 0.6 percent alkali, expressed as Na_2O equivalent. Alternatively, pozzolanic additions are known to be beneficial in reducing ASR.

(2) Avoidance of reactive aggregates. Some types, such as granite or limestone, are unlikely to be reactive. In areas of doubt, expansion tests can be carried out on samples of concrete made from the aggregates in question placed in water at 38℃ for some weeks.

(3) Exclusion of water. Most structures affected to date have been severely exposed, for example, concrete bridges or buildings near the sea. The risk to interior concrete is very small except in swimming pools or other damp areas. Once attack has commenced, it may be very difficult to exclude further water from affected structures.

参 考 译 文

第5课 混 凝 土

1. 概述

混凝土一般是指水泥、集料与水的混合物。加入搅拌机的其他材料被称为"外加剂"。水泥基混凝土材料具有下列优势：

(1) 成本低。

(2) 应用的灵活性强——例如，砂浆、混凝土、水泥浆等。

(3) 可得到各种各样的成品。

(4) 抗压强度高。

(5) 保护埋置的钢筋。

下面是一些缺点或问题：

(1) 较低的抗拉强度/脆性。

(2) 相当高的密度（尽管也有用低密度型的）。

(3) 易受冻/化学劣化影响（取决于其类型）。

2. 水泥

最重要的是波特兰水泥，之所以这么叫是因为用这些水泥制成的混凝土的外形与波特兰石非常相似。

波特兰水泥是水硬性的，即只能通过水的作用才能固结变硬。

1) 波特兰水泥的生产

波特兰水泥的制造方法是将烧窑内磨碎的黏土或页岩以及白垩岩或石灰石加热至高温状态——1500℃左右，以便于实现化合作用；将5%的石膏（硫酸钙）加入到生成的熔渣中以防止瞬时凝固，而且最后阶段还要磨成超细粉体。

2) 波特兰水泥的成分

尽管类似的灰色粉末也能够具有高的均匀度，但重要的是波特兰水泥是一种由黏土或页岩中含有的矿物质与构成石灰石或白垩岩的碳酸钙组成的复杂混合物。下面列出了主要成分及其常用缩写。

(1) 黏土或页岩。

SiO_2 二氧化硅（氧化硅），缩写为 S。

Fe_2O_3 三氧化二铁（氧化铁），缩写为 F。

Al_2O_3 三氧化二铝（氧化铝），缩写为 A。

(2) 石灰石或白垩岩。

$CaCO_3$ 碳酸钙（加热的碳酸钙产生 CaO，即生石灰），缩写为 C。

对水泥熟料的化学分析显示其主要有四种成分,即硅酸二钙(C_2S)、硅酸三钙(C_3S)、铝酸三钙(C_3A),以及铁铝酸四钙(C_4AF)。

波特兰水泥的特性会随着四种成分的比例变化而发生巨大的变化,这体现出了这些成分的性能存在较大的差别。在生产水泥的时候,必须仔细控制黏土与白垩岩的比例(通常重量比大约为1:4),因为这一比例的一个微小变化都会导致硅酸二钙与硅酸三钙的配比发生较大变化。如果白垩岩的比例较大,则后者的配比会更多,因为白垩岩中石灰含量大。这样,会导致水泥的早期强度发展更快。铝酸三钙和铁铝酸四钙对波特兰水泥的长期强度或耐久性的贡献不大,但是很难去除,而且不管怎样其在生产过程中都是有用的。它们通常被用作熔剂以帮助形成硅酸盐化合物,否则这一过程需要更高的窑炉温度。

水泥的特性不仅取决于其成分,还取决于其细度。这是由于水泥颗粒在水里的溶解度较低。这可以通过将少量的水泥与水在烧杯中混合,静置几个小时并偶尔搅拌一次得到证明。

普通波特兰水泥(OPC)是最适合用于加固的水泥;同时,它是最便宜的水泥,而且它将合理的硬化速率与适当的热功率相结合。然而,还可用其他类型的水泥,如快硬波特兰水泥、抗硫酸盐波特兰水泥、低热波特兰水泥、高炉矿渣波特兰水泥及水泥中的粉煤灰(PFA)。

3. 混凝土的集料

出于下列原因,将集料用于混凝土中:

(1) 可以大大降低成本。

(2) 可以降低每单位体积混凝土的热功率,从而降低热应力。

(3) 可以降低混凝土的收缩。

(4) 有助于生产具有良好塑性的混凝土。

还有一些其他的原因使用"特殊的"集料,例如:

(1) 使用低密度混凝土(2000kg/m^3或更小)以降低基础荷载,提高保温能力及降低热惯性。

(2) 根据需要(为屏蔽辐射)可使用高密度混凝土(2600kg/m^3或更大)。

(3) 将耐磨损混凝土用于楼面(花岗岩集料)。

(4) 改进耐火性(石灰岩,轻集料)。

(5) 使用装饰性集料——例如,使用不同颜色的花岗岩石子,可以呈现出外露集料饰面效果。

4. 混凝土用水

用水是将水泥从粉末状变成具有强度和耐久性的硬化材料的化学手段。因此,对水的质量有一些要求而且必须考虑下列两个主要方面:

1) 有机污染

如果水与有机物(如植被)接触,就会受到有机酸的污染。这些有机酸会增大湿性混凝土的pH(酸性)而降低水化速度。如果接触藻类,水会呈现绿色,会导致夹杂空气而强度降低。

2) 易溶盐

海水中可能包括易溶盐,例如能与水化水泥发生反应的硫酸盐,或是能加快水化作用并增加埋置钢筋腐蚀风险的氯化物。如果出现氯化物,就应将水的含盐量加到集料的含盐量中去,以获得二者组合的总数,之后还应将其与允许值作比较。

一般而言,干净的流动水或饮用水均可用于生产混凝土,但是,如果有疑问,也可以依据需要的规格制作混凝土立方体并与用蒸馏水制作的类似混凝土立方体做对比检验。

5. 混凝土的配合比原则

已经考虑了水泥和集料的特性,还有必要确定混凝土的重要特性并检验材料类型与比例对它们的影响。

了解拌和参数如何影响混凝土特性的最重要因素就是水灰比。

$$水灰比 = \frac{混凝土试样中水的质量}{试样中水泥的质量}$$

应该注意的是自由水灰比是最重要的比率,因为它能够影响混凝土的强度与耐久。自由水灰比主要取决于混合物中自由水的含量,即集料吸附的水忽略不计。

另一个常用的术语是集料与水泥(质量)比。

$$集料与水泥比 = \frac{混凝土试样中集料的质量}{混凝土试样中水泥的质量}$$

对于大多数的混凝土混合物而言,集料与水泥(质量)比应在 4~10 之间;如果这一比值较小,说明混合物的水泥富余且造价高;如果这一比值较大,说明混合物的水泥偏少但便宜。尽管通常是依据骨料(砂子)与水泥之比规定混凝土的配比,但是现在更普遍的是用每立方米的水泥用量来规定混凝土拌合物的富余程度。例如,集料与水泥(质量)比为 4 是指水泥含量大约为 $450 kg/m^3$;而集料与水泥(质量)比为 10 是指水泥用量大约为 $200 kg/m^3$。用水泥用量表示拌合物富余程度的一个优势是比较不同富余程度的混凝土的成本相对容易,水泥是其中造价最高的一项。

Lesson 6
Introduction to Mechanics of Materials

Mechanics of materials is a branch of applied mechanics that deals with the behavior of solid bodies subjected to various types of loading①. It is a field of study that is known by a variety of names, including "strength of materials" and "mechanics of deformable bodies". The solid bodies considered in this lesson include axially-loaded bars, shafts, beams, and columns, as well as structures that are assemblies of these components. Usually the objective of our analysis will be the determination of the stresses, strains, and deformations produced by the loads; if these quantities can be found for all values of load up to the failure load, then we will have obtained a complete picture of the mechanical behavior of the body.

Theoretical analyses and experimental results have equally important roles in the study of mechanics of materials. On many occasions we will make logical derivations to obtain formulas and equations for predicting mechanical behavior, but at the same time we must recognize that these formulas cannot be used in a realistic way unless certain properties of the material are known. These properties are available to us only after suitable experiments have been made in the laboratory. Also, many problems of importance in engineering cannot be handled efficiently by theoretical means, and experimental measurements become a practical necessity. The historical development of mechanics of materials is a fascinating blend of both theory and experiment, with experiments pointing the way to useful results in some instances and with theory doing so in others②. Such famous men as Leonardo da Vinci (1452—1519) and Galileo Galilei (1564—1642) made experiments to determine the strength of wires, bars, and beams, although they did not develop any adequate theories (by today's standards) to explain their test results. By contrast, the famous mathematician Leonhard Euler (1707—1783) developed the mathematical theory of columns and calculated the critical load of a column in 1744, long before any experimental evidence existed to show the significance of his results. Thus, Euler's theoretical results remained unused for many years, although today they form the basis of column theory.

The importance of combining theoretical derivations with experimentally determined properties of materials will be evident as we proceed with our study of the subject. In this article we will begin by discussing some fundamental concepts, such as stress and strain, and then we will investigate the behavior of simple structural elements subjected to tension, compression, and shear.

Lesson 6 Introduction to Mechanics of Materials

The concepts of stress and strain can be illustrated in an elementary way by considering the extension of a prismatic bar [Fig. 6-1(a)]. A prismatic bar is one that has constant cross section throughout its length and a straight axis. In this illustration the bar is assumed to be loaded at its ends by axial forces P that produce a

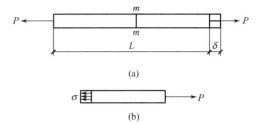

Fig. 6-1 Prismatic bar in tension

uniform stretching, or tension, of the bar. By making an artificial cut (section mm) through the bar at right angles to its axis, we can isolate part of the bar as a free body [Fig. 6-1(b)]. At the right-hand end the tensile force P is applied, and at the other end there are forces representing the action of the removed portion of the bar upon the part that remains. These forces will be continuously distributed over the cross section, analogous to the continuous distribution of hydrostatic pressure over a submerged surface. The intensity of force, that is, the force per unit area, is called the stress and is commonly denoted by the Greek letter σ. Assuming that the stress has a uniform distribution over the cross section [Fig. 6-1(b)], we can readily see that its resultant is equal to the intensity σ times the cross-sectional area A of the bar.③

Furthermore, from the equilibrium of the body shown in Fig. 6-1(b), we can also see that this resultant must be equal in magnitude and opposite in direction to the force P. Hence, we obtain

$$\sigma = \frac{P}{A} \qquad (6-1)$$

as the equation for the uniform stress in a prismatic bar. This equation shows that stress has units of force divided by area—for example, pounds per square inch (psi) or kips per square inch (ksi). When the bar is being stretched by the forces P, as shown in the figure, the resulting stress is a tensile stress; if the forces are reversed in direction, causing the bar to be compressed, they are called compressive stresses.

A necessary condition for Eq. (6-1) to be valid is that the stress σ must be uniform over the cross section of the bar. This condition will be realized if the axial force P acts through the centroid of the cross section, as can be demonstrated by statics. When the load P does not act at the centroid, bending of the bar will result, and a more complicated analysis is necessary. However, it is assumed that all axial forces are applied at the centroid of the cross section unless specifically stated to the contrary. Also, unless stated otherwise, it is generally assumed that the weight of the object itself is neglected, as was done when discussing the bar in Fig. 6-1.

The total elongation of a bar carrying an axial force will be denoted by the Greek letter δ [Fig. 6-1(a)], and the elongation per unit length, or strain, is then determined by the equation

$$\varepsilon = \frac{\delta}{L} \tag{6-2}$$

where L is the total length of the bar. Note that the strain ε is a nondimensional quantity. It can be obtained accurately from Eq. (6-2) as long as the strain is uniform throughout the length of the bar. If the bar is in tension, the strain is a tensile strain, representing an elongation or stretching of the material; if the bar is in compression, the strain is a compressive strain, which means that adjacent cross sections of the bar move closer to one another.

Ⅰ. New Words

1. mechanics *n.* 力学（用作单数），机械学（用作单数），结构，技术
2. deformable *adj.* 可变形的
3. bar *n.* 条，棒，钢筋，障碍，酒吧间
4. shaft *n.* 柄，轴
5. beam *n.* 梁，桁条，束
6. column *n.* 柱，圆柱，专栏，列
7. structure *n.* 结构，构造，建筑物，体系
8. stress *n.* 应力，压力，强调
9. strain *n.* 应变，变形，张力，拉紧
10. derivation *n.* 推导，推论，衍生，起源
11. tension *n.* 张力，拉力；*vt.* 张拉，使拉紧
12. compression *n.* 压缩，受压
13. shear *n.* 剪切，剪力；*vt.* 剪切，修剪
14. prismatic *adj.* 棱形的，棱镜的
15. hydrostatic *adj.* 静水力学的，液体静力学的
16. pressure *n.* 压力，压迫；*vi.* 施加压力
17. resultant *n.* 合力，合成；*adj.* 组合的，合成的
18. equilibrium *n.* 平衡
19. centroid *n.* 形心，矩心
20. statics *n.* 静力学，静态
21. elongation *n.* 伸长，延长

Ⅱ. Phrases and Expressions

1. mechanics of materials 材料力学
2. solid body 固体
3. mechanical behavior 力学性能，机械性能
4. cross section 横截面，横剖面
5. prismatic bar 等截面杆
6. at right angle 直角

7. tensile stress   拉应力
8. compressive stress   压应力
9. nondimensional quantity; dimensionless quantity   无量纲量
10. by contrast   相比之下，对照之下
11. analogous to   类似于

Ⅲ. Notes

① 句中"subjected"是过去分词，与后面的短语一起作后置定语，翻译为"受到各种荷载的……"。

② 句中两个"with"引出各自独立的结构，用"and"连接。"doing"是代动词，指代"pointing"。第二个独立结构的完整形式应为"with theory pointing to useful results in other instances"。

③ 句中的"times"是介词，译为"乘，乘以"。

Ⅳ. Exercises

Fill in the blanks with the information given in the text.

1. Mechanics of materials is a branch of _____ that deals with the behavior of _____ subjected to various types of loading.

2. The _____ of force, that is, the force _____, is called the stress and is commonly denoted by the Greek letter _____.

3. _____, denoted by the Greek letter ε, is the elongation per unit length and is determined by the equation _____. It is a _____ quantity.

Ⅴ. Expanding

Remember the following terms related to the mechanics of materials.
1. basic assumptions; fundamental assumption   基本假设
2. tension, tensile   拉伸
3. compression; compressive   压缩
4. shear   剪切
5. torsion   扭转
6. bending   弯曲
7. elastoplasticity   弹塑性
8. ductile ductility   韧性，延展性
9. brittleness   脆性
10. normal stress   正应力
11. nominal stress   名义应力
12. true stress   真实应力
13. gage length   标准长度
14. percent elongation   延伸率

15. Poisson's ratio; Poisson ratio   泊松比
16. elastic modulus; modulus of elasticity   弹性模量
17. equilibrium equation   平衡方程
18. constitutive relations   本构关系
19. moment of couple   力偶矩
20. vector   矢量
21. torque; twisting moment   扭矩
22. bending moment   弯矩
23. neutral axis   中性轴
24. moment of inertia; inertia moment   惯性矩

Ⅵ. Reading Material

The Tensile Test

The relationship between stress and strain in a particular material is determined by means of a tensile test. A specimen of the material, usually in the form of a round bar, is placed in a testing machine and subjected to tension. The force on the bar and the elongation of the bar are measured as the load is increased. The stress in the bar is found by dividing the force by the cross-sectional area, and the strain is found by dividing the elongation by the length along which the elongation occurs. In this manner a complete stress-strain diagram can be obtained for the material.

The typical shape of the stress-strain diagram for structural steel is shown in Fig. 6-2(a), where the axial strains are plotted on the horizontal axis and the corresponding stresses are given by the ordinates to the curve $OABCDE$. From O to A the stress and strain are directly proportional to one another and the diagram is linear. Beyond point A the linear relationship between stress and strain no longer exists; hence the stress at A is called the proportional limit. For low-carbon (structural) steels, this limit is usually between 30000 psi and 36000 psi, but for high-strength steels it may be much greater. With an increase in loading, the strain increases more rapidly than the stress, until at point B a considerable elongation begins to occur with no appreciable increase in the tensile force. This phenomenon is known as yielding of the material, and the stress at point B is called yield point or yield stress. In the region BC the material is said to have become plastic, and the bar may actually elongate plastically by an amount which is 10 or 15 times the elongation which occurs up to the proportional limit. At point C the material begins to strain harden and to offer additional resistance to increase in load. Thus, with further elongation the stress increases, and it reaches its maximum value, or ultimate stress, at point D. Beyond this point further stretching of the bar is accompanied by a reduction in the load, and fracture of the specimen finally occurs at point E on the diagram.

During elongation of the bar a lateral contraction occurs, resulting in a decrease in the cross-sectional area of the bar. This phenomenon has no effect on the stress-strain diagram up to about point C, but beyond that point the decrease in area will have a noticeable effect upon the calculated value of stress. A pronounced necking of the bar occurs (see Fig. 6-3), and if the actual cross-sectional area at the narrow part of the neck is used in calculating σ, it will be found that the true stress-strain curve follows the dashed line CE'. Whereas the total load the bar can carry does indeed diminish after the ultimate stress is reached (line DE), this reduction is due to the decrease in area and not to a loss in strength of the material itself. The material actually withstands an increase in stress up to the point of failure. For most practical purposes, however, the conventional stress-strain curve $OABCDE$, based upon the original cross-sectional area of the specimen, provides satisfactory information for design purposes.

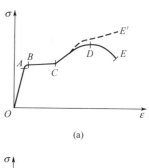

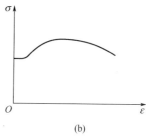

Fig. 6-2 Typical stress-strain curve for structural steel
(a) Pictorial diagram (not to scale);
(b) Diagram to scale

Fig. 6-3 Necking of a bar in tension

The diagram in Fig. 6-2(a) has been drawn to show the general characteristics of the stress-strain curve for steel, but its proportions are not realistic because, as already mentioned, the strain which occurs from B to C may be 15 times as great as the strain occurring from O to A. Also, the strains from C to E are even greater than those from B to C. A diagram drawn in proper proportions is shown in Fig. 6-2(b). In this figure the strains from O to A are so small in comparison to the strains from A to E that they cannot be seen, and the linear part of the diagram appears as a vertical line. The presence of a pronounced yield point followed by large plastic strains is somewhat unique to steel, which is the most common structural metal in use today. Aluminum alloys exhibit a more gradual transition from the linear to the nonlinear region, as shown by the stress-strain diagram in Fig. 6-4. Both steel and many aluminum alloys will undergo large strains before failure and are therefore classified as ductile. On the other hand, materials that are brittle fail at relatively low values of strain (see Fig. 6-5). Examples include ceramics, cast iron, concrete, certain metallic alloys, and glass.

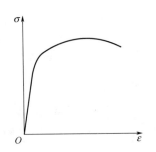
Fig. 6-4 Typical stress-strain curve for structural aluminum alloy

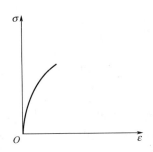
Fig. 6-5 Typical stress-strain curve for a brittle material

Diagrams analogous to those in tension may also be obtained for various materials in compression, and such characteristic stresses as the proportional limit, the yield point, and the ultimate stress can be established. For steel it is found that the proportional limit and the yield stress are about the same in both tension and compression. Of course, for many brittle materials the characteristic stresses in compression are much greater than in tension.

参 考 译 文

第6课 材料力学概述

材料力学是应用力学的一个分支，它讨论固体在承受各种荷载时的性能。它是以"材料强度"和"变形体力学"等不同的名称为人们所熟知的研究领域。本书所讨论的固体包括承受轴向荷载的杆、轴、梁和柱，以及由这些部件装配成的结构物。通常，我们分析的目的是要确定由于荷载所产生的应力、应变和变形；如果对于破坏荷载前的各荷载值都能求得应力、应变和变形的大小，我们就能对物体的力学性能得到完整的概念。

理论分析和试验结果在材料力学研究中具有同等重要的地位。在很多情况下，我们采取逻辑推导得到预测力学性能的公式和方程。但是，必须承认，除非已经知道材料的某些性质，否则这些公式是不能实际应用的，而这些性质只有在实验室做了相应的试验后，我们才会知道。同样，工程中的许多重要问题不能借助于理论手段去有效地处理，因而试验测定就成为实践所必需的。材料力学的历史发展是理论和试验两者最好的结合。在某些情况下，试验能取得有益的成果，而在另一些情况下，理论也会做到这一点。著名的人物如达·芬奇（1452—1519年）和伽利略（1564—1642年）虽然没有按今天的标准提出充分的理论去解释他们的试验结果，但他们用试验确定了金属线、杆和梁的强度。相反，著名数学家欧拉（1707—1783年）早在1744年就提出了柱的数学理论并计算了柱的临界荷载，但很长时间内一直没有试验证明他所得出的结论的重要意义。所以欧拉的理论成果多年未曾得到应用，但是今天它已成为柱的理论基础。

随着对本课程的学习，理论推导与试验确定材料性质相结合的重要性将会愈加明显。本文将先讨论一些基本概念，如应力和应变，然后将研究承受拉伸、压缩和剪切的简单结构部件的性能。

应力和应变的概念可借等截面杆的拉伸［图6-1(a)］进行初步阐述。等截面杆是在整个长度上具有等截面并具有直轴线的杆件。在我们的阐述中，假设轴向力 P 作用于轴的两端，使杆产生均匀的伸长或拉伸。与杆轴线垂直作一人为划切的截面 mm，将杆件分离出一部分成为隔离体［图 6-1(b)］。其右端受拉力 P 的作用，而其另一端的力代表杆件移去部分对留下部分的作用力。这些力沿整个横截面连续分布，类似于浸没面上液体静压力的连续分布。该力的集度，亦即单位面积上的力，称为应力，通常用希腊字母 σ 来表示。假设应力在整个横截面上均匀分布，我们可以很容易地看出其合力等于集度 σ 乘以杆的横截面积 A。

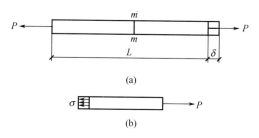

图 6-1 受拉的等截面杆

此外，根据图 6-1(b) 所示物体的平衡，我们还可以看出此合力必定与力 P 的大小相等而方向相反。因此可得等截面杆中均匀应力的方程

$$\sigma = \frac{P}{A} \qquad (6-1)$$

该方程说明应力的单位为力除以面积的单位，例如磅、平方英寸或千磅、平方英寸。如图所示，当杆被力 P 拉伸时，所产生的力为拉应力；如果力反转其方向，使杆压缩，所产生的力则称为压应力。

式(6-1) 能够成立的必要条件是：应力 σ 在杆的整个横截面上必须是均匀的。如果轴向力 P 通过横截面的形心，此条件即会存在，这点可借静力学来证明。当荷载 P 不作用于形心处时，杆将产生弯曲，就需要做较为复杂的分析。但假设所有轴向力均作用于横截面的形心处，改之需特别说明。除非另有说明，一般均假设物体本身重量忽略不计，在讨论图 6-1 中的杆件时，就是这样假定的。

承受轴向力的杆件，其总伸长量用希腊字母 δ 来表示［图 6-1(a)］，单位长度的伸长量或应变，用下列方程来确定

$$\varepsilon = \frac{\delta}{L} \qquad (6-2)$$

式中 L 为杆的总长度。注意，应变 ε 为一无量纲的量。只要应变沿整个杆长是均匀的，就可按式(6-2) 求得其精确值。如果杆件受拉，其应变为拉伸应变，它表示材料伸长或拉伸，如果杆件受压，其应变则为压应变，它表示杆件的相邻横截面彼此移近。

Lesson 7
Site Surveys

1. Summary

A detailed plan of existing ground levels on site is essential if excavation or earth-work filling quantities are to be accurately measured. The most convenient method for surveying and levelling small or moderately sized sites is to use spot levelling on a 20m grid, picking up any sudden changes of level or gradient between these intervals. An instrument man and two chainmen are needed to carry out the work. Ranging poles, two fibreglass tapes, and a number of pegs are required to set out the sight-lines at 20m intervals at right angles from some appropriate baseline. Plotting work is easy, and contours may be reliably interpolated between the 20m interval readings.

Tacheometric work can be faster in the field, but only if a special reading staff is used and the surveyor and chainman are both experienced in the procedure. If tacheometry is tried with a normal staff held vertical (or at right angles to the sight-line of the instrument) a good deal of instrument and calculation work is necessary and, unless the booking down is very clear, plotting work is difficult. Tacheometry is tempting because it offers less resetting of the theodolite: in practice it is not so simple or rapid as spot levelling over a grid.

Aerial surveys may give ground levels up to 2m error, because the camera tends to record the level of the top of the vegetation and not the actual ground level.

Site benchmark levels that will last throughout the job need to be established by the resident engineer. Before going to site he should ascertain what benchmarks were used on the surveys that form the basis of the design, and what levels were taken for them. Hopefully these benchmark levels should still exist: they are usually Ordnance Survey benchmarks on buildings[1], but the original surveyors may also have set up their own benchmarks nearer the site. These (if found) should be rechecked and levels should be brought to convenient benchmarks on the site itself, but in a position unable to be disturbed. At least three such site benchmarks should be established, precisely levelled in. The quickset level is the most convenient and practical to use. It requires only final adjustment of the collimation line for each sighting, which is quickly done. Occasionally the engineer may find to use a three-screw "dumpy" level, which is cheap and robust. Any simpler types of level are not accurate enough for civil engineering work.

The contract drawings will normally include a plan of site levels, but this will probably not have been done in sufficient detail for the proper calculation of excavation and earthwork quantities. The contractor is often responsible for taking all necessary levels, but as the calculation of quantities depends on the levels, the resident engineer's staff should check the levels that the contractor produces. However, to have two surveys of one area is a waste of one party's time. It is best in this case for the resident engineer's staff to undertake the survey, inviting the contractor's site engineer to apply such checks as he wishes.

Responsibility for setting out the works lies with the contractor, but he is entitled to call upon the resident engineer to check that his setting out is correct; although this does not absolve the contractor from holding responsibility for any errors that, even so, escape the notice of the resident engineer and his staff. This is the usual contractual position. But whatever the contract may say, the resident engineer acts unprofessionally if he does not give such reasonable checks as the job demands and his staff can reasonably be expected to undertake, or if he does not immediately inform the contractor on finding some error. The contractor's staff are often hard pressed to set out all necessary levels and sight rails etc. when work is going at full speed, and the resident engineer should answer all calls for checking work. He should also assist in "giving a level" to a foreman on occasion. Both the contractor and resident engineer have a duty to see that the work is properly constructed.

2. Setting out buildings

An accuracy of about 3mm in 30m is desirable; errors of over 5mm in 30m should be rectified. The setting out is done from a suitable baseline by use of a theodolite and steel tape. The appropriate time is when concrete has been placed to column and wall foundations. The baseline, which is either the centre line of the building or a line parallel to it but clear of the building, should have been set out previously by end pegs sited well clear of the work. It is usual to work from coordinates along this baseline from some fixed zero point, and measuring right angle distances out from them. In this way lines of walls and column centres can be marked on the concrete.

Distances have to be measured by steel or fibreglass tape pulled horizontally, so it is a great convenience if the site is level. If not, a plumb-bob has to be used to transfer distances. Distance coordinates along the baseline from the zero peg are set out, using the steel tape and marking a pencil line across the peg. The theodolite is set out over the pencil line, and its position is adjusted laterally so that it transits accurately on the two outermost baseline marks. The plumb-bob on the theodolite gives the mark for the coordinate point, a round-headed nail being inserted on this point. Distances at right angle to the baseline are then set out with theodolite and steel tape. The advantage of this method is that the theodolite can sight down into column bases, which are usually set deeper than the general formation level. For the assistance of bricklayers and formwork carpenters, sight boards can be provided, with the cross-arm fixed a given level above formation level and a saw-cut exactly on the line of sight. A builder's line can then be fixed through such saw-cuts. An alternative to the foregoing is to set out two baselines at right angles to each other and use theodolite right angle settings from these to give centres for such things as column bases.

3. Setting out larger sites

Triangulation from a measured baseline is the usual method adopted, the triangles being as well proportioned as possible. This method is usually better than a lengthy closed traverse when the weather is changeable, as a closed traverse represents more work to be done at one time and may be interrupted by bad weather. If a closed traverse has to be interrupted by bad weather and is left for a day, there is always a danger that one or other of the last two traverse pegs is disturbed by site plant. Even if the pegs have not been disturbed, a large closing error may cause the surveyor to think this has happened, and he will feel it necessary to do the whole job again.

It is worth going to some trouble to find a suitable baseline, which should be level and horizontal. A level road is ideal—if untrafficked! If this is not available it is worth while cutting the grass or removing any humps from a piece of horizontal ground so that the tape may be laid flat and given the standard pull required by means of a spring balance. A new, tested steel tape should be used, the maker's corrections being known and allowed for. A thermometer is also necessary. The temperature should be fairly steady. A quiet, still, cloudy day is best. Accurate measurement of angles with a theodolite is easier than accurate measurement of distances by tape, so it is worth finding a good baseline at the expense, possibly, of not having the best-angled triangles for setting out other points. Preliminary calculations before deciding on the baseline will indicate whether a satisfactory degree of accuracy can be obtained. A 1s theodolite should give an accuracy of ± 1.5mm at 300m.

Distances can also be measured by EDM (electromagnetic distance measurement) equipment, or by electro-optical instruments[②]. The former are used for geodetic surveying over long distances and are not applicable to most site work. The latter can be medium-or short-range instruments, and may be useful for large sites. However, as the average construction site only needs the measurement of one baseline, and most of the rest of work is set out by theodolite and tape work, the expense of this distance-measuring equipment is seldom justified. If a major survey has to be undertaken it is usually more economic to employ an experienced surveying firm. They provide the instruments and experienced surveyors and can get the work done quickly.

When using a theodolite it is best not to resort to more than the standard checks, such as face right, face left. It is often tempting, seeing another previously set-out point, to range in that observation "just as a check". This is likely to lead the engineer into a puzzling conundrum of calculations and trying to decide whether some discrepancy relates to his current survey or the previous one. The trouble is that "check observations" tend to be done with less care than the main survey. Attention should be paid to taking the main theodolite observations with maximum precautions for accuracy: for example, ensuring that the theodolite base is horizontal. Instrument errors due to wear can be a curse, such as when a theodolite gives slightly different readings on a mark for "approach left" and "approach right". Additional precautions are necessary, which must be consistently applied to each reading. Properly maintained instruments and regular recalibration are essential for reducing errors.

When measuring horizontal distances on a steep slope it is a good deal easier to use a light piece of straight timber just over 3m long, on which a 3-m interval is accurately marked, a plumb-bob being suspended from one end mark. Work proceeds downhill, the timber being kept horizontal by means of a builder's level. Pegs are put in at the 3-m intervals, the exact 3-m distance — as shown by the plumb-bob — being marked on each peg.

4. Setting out tunnels

The standard of setting out for tunnels must be high, using carefully calibrated equipment, precise application and double-checking everything. An accurate tunnel baseline is first set out on the surface using the methods described above. Transference of this below ground can be done by direct sighting down a shaft, if the shaft is sufficiently large to allow this without distortion of sight-lines on the theodolite. With smaller shafts plumbing down may be used. A frame is needed either side of the shaft to hold the top ends of the plumb-lines and to allow adjustment to bring them exactly on the baseline. The plumb-line used should be of stainless steel wire, straight and unkinked, and the bob of a special type is held in a bath of oil to damp out any motion. By this means the tunnel line is reproduced at the bottom of the shaft and can be rechecked as the tunnel proceeds.

Many tunnels are nowadays controlled by lasers, the laser gun being set up on a known line parallel to the centre line of the tunnel and aimed at a target. Where tunneling machine is used, the operator can adjust the direction of movement of the machine to keep it on target so that the tunnel is driven in the right direction. For other methods of tunneling, a target may be set in the soffit of the last completed ring or at the tunnel face, the tunnel direction being kept on line by adjusting the excavation and packing out any tunnel rings to keep on the proper line.

Lasers are also used in many other situations, usually for controlling construction rather than for original setting out, as their accuracy for this may not be good enough. The laser beam gives a straight line at whatever slope or level is required and so can be used for aligning forms for road pavements or even laying large pipes to a given fall. For the latter, the laser is positioned at the start of a line of pipes and focused on the required baseline. As each new pipe is fitted into the pipeline a target is placed in the invert of the open end of the pipe, using a spirit-level to find the bottom point, and the pipe is adjusted in line and level until the target falls on the laser beam. Bedding and surround to the pipe are then placed to fix the pipe in position.

Ⅰ. New Words

1. survey *vt.* 勘测，调查，俯瞰
2. moderately *adv.* 适度地，中庸地，有节制地
3. gradient *n.* 梯度，坡度，倾斜度
4. peg *n.* 钉，桩
5. contour *n.* 等高线；轮廓
6. interpolate *vt.* 插（值），增添
7. tacheometric *adj.* 视距仪的，视距测量的
8. surveyor *n.* 测量员，勘测员，检验员
9. tacheometry *n.* 视距测量，视距测量法
10. theodolite *n.* 经纬仪
11. vegetation *n.* 植物，植被，草木
12. benchmark *n.* 基准，基准点，水准点
13. ascertain *vt.* 确定，查明，探知
14. ordnance *n.* 军械，军用器材；军械部门
15. collimation *n.* 瞄准
16. screw *n.* 螺栓，螺杆，螺丝钉
17. dumpy *adj.* 矮胖的，粗短的
18. robust *adj.* 强健的，粗野的
19. absolve *vt.* 免除，赦免，宣告……无罪
20. rectify *vt.* 矫正，纠正
21. triangulation *n.* 三角测量，三角形划分
22. lengthy *adj.* 漫长的，冗长的
23. thermometer *n.* 温度计，体温表
24. geodetic *adj.* 大地测量学的
25. recalibration *n.* 重新校准
26. calibrate *vt.* 校准，校正，调整
27. distortion *n.* 变形，歪曲，失真
28. kink *vt.* 使扭结
29. soffit *n.* 底面，背面；拱腹
30. align *vt.* 使成一行，对准，校直
31. invert *vt.* 转化，颠倒，反转

Ⅱ. Phrases and Expressions

1. Ordnance Survey （英国）国家测绘局，全国地形测量
2. electro-optical instruments 光电仪器

Ⅲ. Notes

① 句中"Ordnance Survey"是指英国国家测绘局（The Great Britain's National Mapping Agency），可简写为 OS。

② 句中的"electromagnetic distance measurement equipment"译为"电磁波测距仪"，"electro-optical instrument"译为"光电仪"。

Ⅳ. Exercises

1. Translate the following words into English or Chinese.

（1）水准测量 _____

（2）水准仪 _____

（3）水准尺 _____

（4）水准点 _____

（5）Optical theodolite _____

（6）Trigonometric leveling _____

（7）air pressure leveling _____

（8）GPS leveling _____

2. Fill in the blanks with the information given in the text.

Distances can also be measured by EDM (_____) equipment, or by _____ instruments.

Ⅴ. Expanding

Remember the following terms related to engineering survey.

1. engineering surveying   工程测量学
2. angular observation   角度测量
3. distance measure   距离测量
4. coordinate measurement   坐标测量
5. horizontal angle observation   水平角测量
6. vertical angle observation   垂直角测量
7. steel tape measuring   钢尺测距
8. stadia measuring   视距法测距
9. line orientation   直线定向
10. system error   系统误差
11. accident error   偶然误差

Ⅵ. Reading Material

Site Drainage

Site drainage and flood protection is important. Difficulty often occurs in draining a site where large-scale earthmoving is taking place. The excavations disturb the natural drainage of the land, and large quantities of mud may be discharged to local watercourses during wet weather. Complaints then arise from riparian owners and water abstractors downstream. If this possibility should occur, the resident engineer should advise the contractor to approach the appropriate drainage authority (the National Rivers Authority in England and Wales) to seek advice on the best course of action to alleviate the problem, such as arranging some form of tank to pond the runoff and allow the heaviest suspended solids to settle out. It is the contractor's responsibility to dewater the site, and this includes the obligation to do so without causing harm or damage to others.

Dewatering can range from simple diversion or piping of ditches, to full-scale 24-hour pumping and groundwater table lowering. It is usual to cut perimeter drains on high ground around all extensive excavations. In dry weather this may seem a waste of time, but once wet weather ensues and the ground becomes saturated, further rain may bring a storm runoff of surprising magnitude. If no protection exists for these occasions extensive damage can be caused to both temporary and permanent works. The resident engineer should assist the contractor to appreciate the danger of flood damage by providing him with data showing possible flood magnitudes. A frequently used precaution is to assume that a flood of magnitude 1 year in 10 (i.e. 10 percent probability) will occur during the course of construction.

The need to dewater an excavation in the British Isles is the rule rather than the exception. Once dewatered an excavation should be kept dewatered. It can be dangerous to repeatedly dewater an excavation during the day and let it fill up overnight; this can cause ground instability, and timbering to excavations may be rendered unsafe. The need for 24-hour pumping should be insisted upon by the resident engineer if he thinks damage or danger could occur from intermittent dewatering. The electric self-priming centrifugal pump is the most reliable for continuous dewatering, having the advantage that it is relatively silent for night operation as compared with petrol or diesel engine driven pumps.

For groundwater lowering, pointed and screened suction pipes are jetted into the ground at intervals around a proposed excavation and are connected to a common header suction pipe leading to a vacuum pump. It may take a week or more before the groundwater is lowered but, when the process works well (as in silt or running sand), the effect is quite remarkable. It permits excavation to proceed with ease in ground that, prior to dewatering, maybe semi-liquid. However, it can be difficult to get the well points jetted down into ground containing cobbles and boulders; and in clays the well points need to be protected by carefully graded filters, or the withdrawal of water may eventually diminish because the well point screens become sealed by clay.

Special precautions must be taken to avoid damage to any adjacent structures when dewatering any excavation or groundwater lowering. In some soils groundwater lowering may cause building foundations to settle, causing considerable damage. The contractor may have to provide an impermeable barrier between the pumped area and nearby structures, monitor water levels, and perhaps provide for recharge of groundwater under structures. A vital precaution is for the resident engineer to record in detail all signs of distress (cracks, tilts etc.) in adjacent structures and take photographs of them, dated and sized, before work starts, in order to provide evidence of the extent of any damage that may occur.

The drainage of clay or clay and silt can present difficulty. The problem is not so much that it cannot be done, but that it can take a long time, perhaps many weeks. Sand drains, such as bored holes filled with fine sand, can be satisfactory as part of the permanent design of the works, but they usually operate too slowly to be of use during construction. If the ground is too soft, any attempt to start excavating it by machine may make matters considerably worse, and end with the machine having to be hauled out. The act of removing overburden may make a soft area even softer as springs and streams, otherwise restrained by the overburden material, break out and change the area to a semi-liquid state. If the resident engineer sees the contractor moving towards these difficulties he should advise him of the possible consequences, and endeavor to give assistance in devising a better approach. A paramount need may be to call in an experienced geotechnical engineer to investigate the problem and give advice as to the best policy to handle the situation.

参 考 译 文

第 7 课　场 地 测 量

1. 概述

如果要精确地测量开挖或回填土方工作量，制订一份现场现有地面基准面的详细（测量）计划至关重要。测量和平整小型或中型场地最方便的方法是在 20m 网格上使用点调平，得到这些间隔间的任何水平或梯度上的突变。这需要一个仪表工和两个丈量员来完成这项工作。需要用测距杆，两个玻璃纤维卷尺和一些定位桩，从与适当的基线相垂直处以 20m 的间隔设置视线。绘制工作很容易，而且等高线上可以以 20m 间隔读数之间可靠地插值。

现场采用视距测量法工作可以更快，但需使用指定的读数人员，并且测量员和丈量员都是在该过程中有经验的人员。如果试图让工作人员保持直立（或与仪器视线成直角）进行视距测量，则需要大量的仪器和计算工作，除非记录得非常清楚，否则绘图工作很困难。视距测量法很吸引人，是因为它极少需要重新设置经纬仪；在实践中，它不会像网格上的点调平那么简单或快速。

航空测量可以使地平面达到 2m 的误差，因为相机总是易于记录植被的顶部平面，而不是实际地平面。

需要由常驻工程师确定整个工作期间的场地基准平面。在进入现场之前，他应该确定在构成设计基础的测量中使用了多少水准点，以及他们采用了什么基准面。希望这些基准平面应该还在：他们通常在建筑物上使用英国国家测绘局的水准点，但是最初的测量人员也可以在现场附近设置自己的水准点。这些水准点（如果发现有）应该重新校核，并且基准面应该为现场本身带来方便的基准，但是应该处在一个不被打扰的位置。至少应建立三个这样的场地水准点，精确调平。快速调平是最方便和实用的。它仅需要对每次瞄准的瞄准线进行最终调整，这可以很快完成。偶尔，可能会发现工程师使用三螺杆"水平仪"，该水平仪便宜且耐用。任何简单型的水平仪在土木工程工作中都不够准确。

合同图纸通常包括现场平整计划，但对于挖方和土方量的合理计算，可能没有更多的细节之处。承包商通常负责获取所有必需的基准面，但由于工程量的计算取决于基准面的位置，驻地的工程师人员应校核承包商制定的基准面。然而，对一个区域进行两次测量是在浪费各方的时间。在这种情况下，最好由驻地工程师人员进行测量，邀请承包商的现场工程师按照他的要求进行校核。

承包商负责布置测量工作，但他有权要求驻地工程师检查他的布置工作是否正确；虽然这并不能免除承包商对任何错误所应承担的责任，即使如此，也不能无视驻地工程师及其工作人员的（检查）通知。这是通常的合同立场。但无论合同如何，如果常驻工程师不能按工作需要给予这样合理的检查以及他的工作人员不能按预期进行，或者他发现的错误没有立即通知承包商，那么该工程师就表现得太不专业了。承包商的员工通常很难设置所有必需的水平仪和照准轨等。当工作全速进行时，驻地工程师应该回复所有检查工作的要求。有时他也可以帮助工头"给出基准面"。承包商和驻地工程师都有责任监督工程建设正确进行。

2. 建筑物的测定

在 30m 内精度约为 3mm 是理想的；30m 内误差超过 5mm 就应该纠正。通过使用经纬仪和钢尺可以从合适的基线进行测定。测定的合适时间是将混凝土浇筑于柱和墙基础时。基线，或者是建筑物的中心线，或者是平行于建筑物的线，但不是建筑物本身，应该事先通过端桩定位好。通常工作以从某个固定的零点开始沿着该基线的坐标为基准，并且从中测量直角距离。以这种方式，可以在混凝土上标出墙和柱的中心线。

距离必须通过水平拉拽钢或玻璃纤维卷尺来测量，如果场地平坦，这就非常方便。如果场地不平，则必须使用铅锤来测量距离。使用钢尺并标出穿过点位桩的铅垂线，从零点定位桩沿着基线的距离坐标就可测定出来。经纬仪放置在铅垂线上，其位置经横向调整，使其在两个最外侧基线标记上精确地传输。经纬仪上的铅锤给出了坐标点的标记，在该点上插上圆头钉。然后用经纬仪和钢尺确定与基线成直角的距离。这种方法的优点是经纬仪可以向下看到柱基，通常设置比一般平整面标高更深。为了获得瓦匠和木工的帮助，可以提供观察板，其中横臂固定在平整面标高以上的给定平面上，并精确地固定在视线的切谷上。通过这种切谷可以固定建筑工人的基准线。上述方案的替代方案是设置两个彼此成直角的基线，并使用来自这两个基线的经纬仪直角设置给出诸如柱基等构件的中心。

3. 更大场地的测定

基于一条已知基线进行三角测量是通常采用的方法，三角形要尽可能地成比例。当天气经常变化时，该方法通常比冗长的闭合导线更好，因为闭合导线要一次完成更多的工作，并且可能被恶劣天气中断。如果闭合导线因天气恶劣而不得不中断并且要持续一天的话，这总是个危险，即最后两个导线挂钩中的一个或另一个被现场设备扰动。即使导线挂钩没有被扰动，一个大的闭合误差也可能导致测量员认为已经发生扰动，他会觉得有必要重新做整个工作。

通过这些问题来找到一条合适的基线是值得的，这条基线应该是水平面和水平线。一条水平的路面是理想的选择——如果不行车！如果该路面不可用，则需要从一块水平的地面上割掉草皮或去除任何隆起，这样钢尺可以放平并通过弹簧秤给出所需的标准拉力。应使用新的检测过的钢尺，制造商的修正是已知的并经过允许的。温度计也是必要的，温度应该相当稳定。用经纬仪精确测量角度比用钢尺精确测量距离更容易，所以寻找一条好基线是值得的，代价可能是没有用于测定其他点的最佳角度的三角形。在确定基线之前的初步计算会表明是否可以获得令人满意的精度。1s 经纬仪应在 300m 内给出 ±1.5mm 的精度。

距离也可以通过 EDM（电磁波测距）设备或电光学仪器测量。前者用于长距离的大地测量，不适用于大多数的现场工作。后者可以是中程或短程仪器，并且可用于大型场地。然而，由于一般的施工现场只需要测量一条基线，其余的大部分工作由经纬仪和钢尺来测定，因此这种测距仪器的使用通常都不太合理。如果必须进行重大测定，通常雇用经验丰富的测绘公司会更经济。他们能提供仪器和经验丰富的测量师，可以快速完成工作。

当使用经纬仪时，最好不要进行多于标准的检查，如盘右，盘左。在看到另一个先前测定点时，人们往往忍不住将检查归于"只是个检查"而已。这可能会导致工程师陷入令人困惑的计算难题，并试图确定这些差异是否与他当前的测量或之前的测量有关。麻烦的是，"检验观测"更倾向于对主要测定很少关注。应注意使用带有最大准确性预防措施的主经纬仪观测，例如，确保经纬仪基座是水平的。由于磨损导致的仪器误差可能是祸因，例如当经纬仪对"向左"和"向右"的标记给出稍微不同的读数时，需要额外的预防措施，这必须一致地应用于每个读数。正确维护仪器和定期重新校准对于减少误差至关重要。

当在陡峭的斜坡上进行水平测距时，用一块较轻的刚好超过 3m 长的直木方很容易实现，其上精确地标记了 3m 的间隔，铅锤悬挂在一个标记端。工作沿着下坡进行，木方通过建筑工的调平保持水平。定位桩间隔 3m 放置，精确的 3m 距离——如铅锤所示——被标记在每个定位桩上。

4. 隧道的测定

隧道的测定标准必须很高，应使用精心校准的设备，精确应用并反复检查每一处。首先使用上述方法在表面上测定准确的隧道基线。如果轴足够大能允许其在经纬仪上的视线不变形，则可以通过直接向下看轴来进行地面下方的传输。管道下方可以使用较小的轴。轴的任一侧都需要支架来支撑铅垂线的顶端并且允许调节，以将它们精确地放到基线上。所使用的铅垂线应该是不锈钢丝，直的并且是松开的，特殊类型的铅锤应保持在油浴中，以减少任何摆动。通过这种方式，隧道基线会在轴的底部再现并随着隧道工程的推进可重新检查。

许多隧道现在由激光控制，激光枪设置在与隧道的中心线平行的已知线上并用来瞄准目标。在使用隧道挖掘机的地方，操作者可以调节仪器的运动方向将其保持在目标处，使得隧道挖掘工作在正确的方向上进行。对于其他的隧道挖掘方法，可以在最后完成的环路的拱腹内或在隧道表面处设定目标，通过调整挖掘来保持隧道方向在线上，并且填塞任何隧道环路以保持在正确的基线上。

激光仪也用在许多其他情况，通常用于控制施工而不是用于初始测定，因为用于初始测定的精度不够好。激光束在任何斜面或水平面上都能产生直线，因此可用于对准路面的形状甚至将大型管道铺设到给定的落差处。对于后者，激光定位在管线的开始处并且聚焦在所需的基线上。将每个新管道装配到管线中时，目标被放置在管道的开口端的倒拱处，使用水平仪来找到底点，并且管道在管线和水平面上进行调整，直到目标落在激光束上。将基床和管道的环绕件布置好后将管道就位固定。

Lesson 8
Soil Mechanics

1. Soils

There are two broad classes of soils—coarse-grained soils and fine-grained soils.

Coarse-grained soils include gravel and sand, which consist of relatively large particles visible to the naked eye; fine-grained soils, such as silt and clay, consist of much smaller particles. The American Society for Testing and Materials (ASTM[①]) Unified Soil Classification System further divides gravels, sand, silts, and clays into soil types based on physical composition and characteristics. See Table 8-1 that follows.

Table 8-1 Soil classification

Soil Classification[*1]		Symbol	Description	Presumptive Bearing Capacity[*2]		Susceptibility to Frost Action	Permeability & Drainage
				psf[*3]	kPa		
Gravels 6.4—76.2mm	Clean gravels	GW	Well-graded gravel	10000	479	None	Excellent
		GP	Poorly graded gravel	10000	479	None	Excellent
	Gravels w/fines	GM	Gravel	5000	239	Slight	Poor
		GC	Clayey gravel	4000	192	Slight	Poor
Sands 0.05—6.4mm	Clean sands	SW	Well-graded sand	7500	359	None	Excellent
		SP	Poorly graded sand	6000	287	None	Excellent
	Sands w/fines	SM	Silty sand	4000	192	Slight	Fair
		SC	Clayey sand	4000	192	Medium	Poor
Silts 0.002—0.05mm	LL>50[*4]	ML	Inorganic silt	2000	96	Very high	Poor
		CL	Inorganic clay	2000	96	Medium	Impervious

(Continued)

Soil Classification[1]		Symbol	Description	Presumptive Bearing Capacity[2]		Susceptibility to Frost Action	Permeability & Drainage
				psf[3]	kPa		
Clays < 0.002mm	LL<50[4]	OL	Organic silt-clay			Very poor	High	Impervious
		MH	Elastic inorganic silt	2000	96	Very high	Poor
		CH	Plastic inorganic clay	2000	96	Medium	Impervious
		OH	Organic clay and silt			Very poor	Medium	Impervious
Highly organic soils		Pt	Peat		Unsuitable	Slight	Poor

* 1　Based on the ASTM Unified Soil Classification System.
* 2　Consult a geotechnical engineer and the building code for allowable bearing capacities.
* 3　1 psf=0.0479kPa②.
* 4　LL= liquid limit: the water content, expressed as a percentage of dry weight, at which a soil passes from a plastic to a liquid state.

　　The soil underlying a building site may actually consist of superimposed layers, each of which contains a mix of soil types, developed by weathering or deposition. To depict this succession of layers or strata called horizons, geotechnical engineers draw a soil profile, a diagram (Fig. 8 - 1) of a vertical section of soil from the ground surface to the underlying material, using information collected from a test pit or boring.

　　The integrity of a building structure depends ultimately on the stability and strength under loading of the soil or rock underlying the foundation. The stratification, composition, and density of the soil bed, variations in particle size, and the presence or absence of groundwater are all critical factors in determining the suitability of a soil as a foundation material. When designing anything other than a single-family dwelling, it is advisable to have a geotechnical engineer undertake a subsurface investigation.

　　A subsurface investigation (CSI Master Format ^{TM}02 32 00) involves the analysis and testing of soil disclosed by excavation of a test pit up to 10 feet (3 meters) deep or by deeper test borings in order to understand the structure of the soil, its shear resistance and compressive strength, its water content and permeability, and the expected extent and rate of consolidation under loading. From this information, the geotechnical engineer is able to gauge the anticipated total and differential settlement under loading by a proposed foundation system.

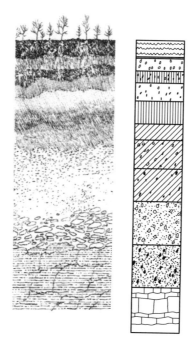

Fig. 8 – 1   A soil profile

2. Soils mechanics

The allowable bearing capacity of a soil is the maximum unit pressure a foundation is permitted to impose vertically or laterally on the soil mass. In the absence of geotechnical investigation and testing, building codes may permit the use of conservative load-bearing values for various soil classifications. While high-bearing-capacity soils present few problems, low-bearing-capacity soils may dictate the use of a certain type of foundation and load distribution pattern, and ultimately, the form and layout of a building.

Density is a critical factor in determining the bearing capacity of granular soils. The Standard Penetration Test measures the density of granular soils and the consistency of some clays at the bottom of a hammer to advance a standard soil sampler. In some cases, compaction, by means of rolling, tamping, or soaking to achieve optimum moisture content, can increase the density of a soil bed.

Coarse-grained soils have a relatively low percentage of void spaces and are more stable as a foundation material than silt or clay. Clay soils, in particular, tend to be unstable because they shrink and swell considerably with changes in moisture content. Unstable soils may render a site unbuildable unless an elaborately engineered and expensive foundation system is put in place.

The shearing strength of a soil is a measure of its ability to resist displacement when an external force is applied, due largely to the combined effects of cohesion and internal friction. On sloping sites, as well as during the excavation of a flat site, unconfined soil has the potential to displace laterally. Cohesive soils, such as clay, retain their strength when unconfined; granular soils, such as gravel, sand, or some silts, require a confining force for their shear resistance and have a relatively shallow angle of repose (Fig. 8-2).

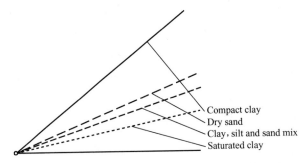

Fig. 8-2 Angle of repose for bare soil embankments

The water table is the level beneath which the soil is saturated with groundwater. Some building sites are subject to seasonal fluctuations in the level of groundwater. Any groundwater present must be drained away from a foundation system to avoid reducing the bearing capacity of the soil and to minimize the possibility of water leaking into a basement. Coarse-grained soils are more permeable and drain better than fine-grained soils, and are less susceptible to frost action.

Ⅰ. New Words

1. gravel *n.* 砾，砾石
2. silt *n.* 淤泥，泥沙
3. superimposed *adj.* 叠置的，上层遗留的
4. weathering *n.* 风化（作用），侵蚀
5. deposition *n.* 沉积（作用）
6. stratification *n.* 成层，层理，地层
7. permeability *n.* 渗透，渗透性
8. gauge *vt.* 测量，测定，估计，判定
9. granular *adj.* 颗粒的，颗粒状的
10. penetration *n.* 穿入，穿透，穿透深度，穿透能力
11. sampler *n.* 取样器，抽样器
12. compaction *n.* 压实，压紧，紧密
13. cohesion *n.* 黏聚力，凝聚力
14. excavation *n.* 挖掘，挖土，开凿
15. unconfined *adj.* 无侧限的，自由的，松散的
16. saturate *vt.* 使饱和，使达到饱和，使充满
17. fluctuation *n.* 波动，起伏，不稳定

Ⅱ. Phrases and Expressions

1. coarse-grained soil   粗粒土

2. fine-grained soil 细粒土
3. liquid limit 液限
4. soil profile 土层剖面
5. geotechnical investigation 岩土工程勘探
6. shear resistance 抗剪承载力
7. compressive strength 抗压强度
8. water content 含水量
9. rate of consolidation 固结速率
10. differential settlement 不均匀沉降
11. allowable bearing capacity 容许承载力
12. granular soil 粒状土
13. standard penetration test 标准贯入试验
14. internal friction 内摩擦
15. groutd water table 地下水位，潜水面
16. frost action 冻裂作用，冰冻作用
17. angle of repose 休止角

Ⅲ. Notes

① "ASTM" 是美国材料与试验协会（American Society for Testing and Materials）的英文缩写。它的前身是国际材料试验协会（International Association for Testing Materials，IATM）。19 世纪 80 年代，为解决采购商与供货商在购销工业材料过程中产生的意见和分歧，有人提出建立技术委员会制度，由技术委员会组织各方面的代表参加技术座谈会，讨论解决有关材料规范、试验程序等方面的争议问题。IATM 首次会议于 1882 年在欧洲召开，并组建了工作委员会。

② psf，计量单位（磅/平方英尺），英文全称为 "pounds per square foot"，1psf＝0.0479kPa。

Ⅳ. Exercises

Fill in the following blanks with right words.

1. There are two broad classes of soils— _____ and _____ .

2. _____ : the water content, expressed as a percentage of dry weight, at which a soil passes from a plastic to a liquid state.

3. The shearing strength of a soil is a measure of its ability to resist displacement when an external force is applied, due largely to the combined effects of _____ and _____ .

4. _____ is the level beneath which the soils is saturated with groundwater.

Ⅴ. Expanding

Know about the following terms related to the soil mechanics.

1. unit weight 容重
2. saturated unit weight 饱和容重
3. buoyant unit weight 浮容重
4. specific gravity 比重
5. bulk density 体积密度
6. dry density 干密度
7. buoyant density 浮密度
8. relative density 相对密度
9. void ratio 孔隙比
10. porosity 孔隙率
11. degree of saturation 饱和度
12. liquidity index 液性指数
13. plasticity index 塑性指数
14. plastic limit 塑限
15. shrinkage limit 缩限

Ⅵ. Reading Material

Three Types of Soil Particles Sized from Biggest to Smallest

There are three types of soil particles: sand, silt and clay. Most soils are made up of a combination of sand, silt and clay particles. The ratio of these particles in any given soil sample identifies it as one of the three main types of soil: sandy, loam or clay.

1. Sand particles

Sand is the biggest soil particle. If blown up to an easily visible size, compared to other soil particles, a sand particle would be the size of a basketball. Soils classified as sandy typically contain 80 to 100 percent sand, zero to 10 percent silt and zero to 10 percent clay by volume. Sandy soils, because of the large size of their particles, do not hold water well and have low nutrient value. Grasp a handful of sandy soil in your hand and it will crumble easily when you let it go.

2. Silt particles

Silt is the next largest soil particle. If blown up to an easily visible size, compared to other soil particles, a silt particle would be the size of a baseball. Soils classified as loam have the largest amount of silt particles and typically contain 25 to 50 percent sand, 30 to 50 percent silt and 10 to 30 percent clay by volume. High-silt-content soils are often found along riverbanks. Silt soils feel smooth when wet and powdery when dry. Grasp a handful of loam or silty soil in your hand and it will retain its shape when you let it go if it is moist but crumble away when dry.

3. Clay particles

Clay is the smallest soil particle. If blown up to an easily visible size, compared to other soil particles, a clay particle would be the size of a golf ball. Soils classified as clay typically contain zero to 45 percent sand, zero to 45 percent silt and 50 to 100 percent clay by volume. Because of the small size of clay particles, clay traps water and air, making it difficult for plants to grow. Grasp a handful of clay soil in your hand and it will retain the shape of your hand, like modeling clay, when you let it go. When dry, it is often too hard to easily penetrate with a shovel.

参考译文

第 8 课 土 力 学

1. 土

土分为两大类——粗粒土和细粒土。

粗粒土包括砾粒和砂粒，是由肉眼能够看到的相对较大的颗粒组成；细粒土是由非常小的颗粒构成的，例如淤泥及黏粒。"美国材料与试验协会统一标准土壤分类体系"根据土的物理组成及特性进一步将砾粒、砂粒、淤泥和黏粒分成各种土的类型，详见表 8-1。

表 8-1 土的分类

土的分类[1]		符号	描述	假定承载力[2]		对冰冻作用的敏感性	渗透性与渗流
				psf[3]	kPa		
砾类土 6.4~76.2mm	纯砾石	GW	级配优良的砾石	10000	479	无	极好
		GP	级配不良的砾石	10000	479	无	极好
	含细粒类砾石	GM	砾石	5000	239	轻微	差
		GC	黏土质砾石	4000	192	轻微	差
砂类土 0.05~6.4mm	纯砂土	SW	级配优良的砂土	7500	359	无	极好
		SP	级配不良的砂土	6000	287	无	极好
	含细粒类砂土	SM	细泥砂	4000	192	轻微	中等
		SC	黏土质砂	4000	192	中等	差
泥浆 0.002~0.05mm	LL>50[4]	ML	无机泥浆	2000	96	非常高	差
		CL	无机黏土	2000	96	中等	不受影响

(续)

土的分类*1		符号	描述	假定承载力*2		对冰冻作用的敏感性	渗透性与渗流
				psf*3	kPa		
黏土<0.002mm	LL<50*4	OL	有机粉质黏土		非常差	高	不受影响
		MH	弹性无机泥浆	2000	96	非常高	差
		CH	塑性无机黏土	2000	96	中等	不受影响
		OH	有机黏土和粉砂		非常差	中等	不受影响
高度有机土		Pt	腐殖土		不合适	轻微	差

*1 依据"美国材料与试验协会统一标准土壤分类体系"。

*2 在岩土工程师和建筑规范中查阅容许承载力。

*3 1psf=0.0479kPa。

*4 LL=液限：土从可塑状态变至流动状态的含水量，以干重量的百分比表示。

建筑场地下面的土可能是由许多叠加层构成，每个叠加层是风化作用或沉积作用形成的各种类型的土的混合。为了描述这些叠加层或被称为地平线的地层的层理，岩土工程师利用从探井或钻探中收集的信息绘制了一个土层剖面，一个从地表面到底层的材料之间的土垂直剖面图（图 8-1）。

从根本上说，建筑结构的整体性取决于基础下面的岩土在荷载作用下的稳定性与强度。土床的层理、成分和密度，颗粒大小的变化以及地下水的存在与否等均是决定土是否适合做地基材料的关键因素。在设计除了独户住房之外的建筑时，找一位岩土工程师进行地质勘测是明智的。

地质勘探（CSI Master Format ™02 32 00）是指分析并测试挖至 10 英尺（3m）深的探井或更深的勘探钻井的土，以便于了解土的结构、剪切强度及抗压强度、含水量及渗透性，以及在荷载作用下土的预期固结度和固结率。根据这些信息，岩土工程师能够估计出在拟建基础体系施加的荷载作用下可能的总沉降量以及不均匀沉降。

2. 土力学

土的容许承载力是指容许基础施加在土体上的竖向或水平向的最大单位压力。当缺乏岩土工程勘察和测试时，建筑规范允许取各种类型土的承载力的保守值。承载力低的土可能决定基础类型的选用和荷载的分布形式，最终可能决定建筑物的形式与布局；而承载力高的土几乎没有这些问题。

密度是决定颗粒土承载力的一个关键因素。标准贯入度试验是通过锤子底部贯入标准土壤取样器的深度来测量颗粒土的密度和一些黏土的稠度。在某些情况下，通过滚压、夯实或通过浸泡使土达到最佳含水量等方式的压实，能够提高土床的密度。

粗粒土具有相对较低的孔隙率，其作为地基材料要比淤泥和黏土稳定。尤其是黏质土，会随着含水量的变化而显著收缩或膨胀，因此更加不稳定。不稳定的土可能导致场地不适合建造房屋，除非在场地建造一个精心设计的、昂贵的地基体系。

由于土的黏聚力和内摩擦的共同作用，在施加外力时土会产生位移，其抗剪强度即是衡量抵抗这种位移的能力。在倾斜地基上以及在开挖平坦地基的过程中，松散的土可能会发生横向位移。黏性土，如黏土，在松散状态下可以保持其强度；而颗粒土，如砾石、砂或淤泥，需要一个约束力才能实现其抗剪能力，且休止角较小（图8-2）。

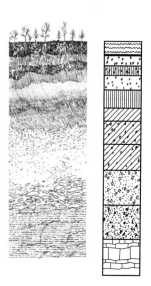

图8-1 土的剖面图

地下水位是指地下水使其下的土饱和时的水位。一些建筑工地受地下水位季节性波动的影响。必须将地基系统中的地下水排走以避免土承载力的降低，以及把水渗进地下室的可能性降到最低。与细粒土相比，粗粒土的渗透性及排水能力更好，而且更不易受霜冻的影响。

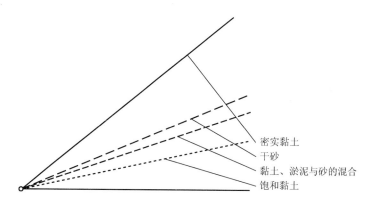

密实黏土
干砂
黏土、淤泥与砂的混合
饱和黏土

图8-2 裸土路堤的休止角

Lesson 9

Shallow Foundations Forms

The nature of the foundation that is appropriate in a particular situation depends on a number of factors. The extent of the loading and the way in which it is applied to the foundation dictates the design, together with the nature of the supporting strata.

In general, the foundation solutions available for the construction of dwellings are restricted to simple forms; these are normally defined within the classifications of shallow foundations and deep foundations. We will discuss about shallow foundations.

A range of shallow foundation forms are available and the nature of the selected construction form, the likely loads and the bearing capacity of the ground will all be considered when making a selection.

1. Strip foundations

The most common form of foundation in use is the simple mass concrete strip foundation. Used where continuous lengths of wall are to be constructed, the strip foundation consists of a strip of mass concrete of sufficient width to spread the loads imposed on it over an area of the ground which ensures that overstressing does not take place. The distribution of the compressive loading within the concrete of the strip tends to fall within a zone defined by lines extending at 45° from the base of the wall. Hence, to avoid shear failure along these lines, the depth of the foundation must be such that the effective width of the foundation base falls within this zone (Fig. 9-1).

The loads exerted on the central zone of the foundation are distributed within the concrete at a 45°angle [Fig. 9-1(a)]. This results in the creation of a defined area of loading at the point of contact with the ground. The loads are resisted by the reaction force from the ground. If the width of foundation is such that this reaction is applied outside the zone of applied loading shear failure may result. In order to avoid the risk of shear failure [Fig. 9-1(b)] the foundation is designed so that the foundation does not extend outside the zone of downward loading. The width of foundation is dictated by the strength of the soil and the extent of the applied load. (Area is adjusted to reduce pressure.) As the width increases, so the depth of the foundation is increased so that there is no chance of shear failure.

Strip foundations are intended for use where the loads from the building are relatively modest, as in the case of house building, and are distributed uniformly (Fig. 9-2). Since the external walls of dwellings transfer the structural loads to the foundations they are

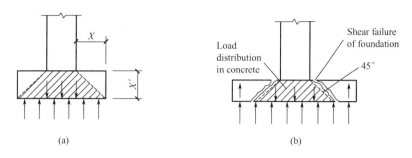

Fig. 9 – 1 Foundation design to avoid shear failure

(a) $X' \geqslant X$ to avoid risk of shear failure; (b) Reaction from ground support

generally uniform in distribution. In some forms of construction the loads are transferred as localized or point loads. In such instances the loads are catered for by the adoption of isolated foundations at each point of load. This is rarely encountered in house building.

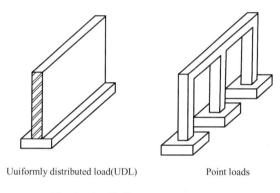

Uuiformly distributed load(UDL)　　　Point loads

Fig. 9 – 2 Uniform and point loads

The traditional form of most house construction allows the use of shallow strip foundations in most instances. The width of the foundation depends on the safe bearing capacity of the soil and the level of loading applied. The width of the foundation is adjusted to provide a sufficient area of contact to prevent overstressing of the soil. The depth will be related to the depth of appropriate bearing strata, but will generally be in excess of 1metre (Fig. 9 – 3) so as to avoid the danger of frost-related heave and shrinkage of the soil close to the surface.

Excavation for the placement of foundations takes place after the building has been set out to indicate the positions of trenches using profile boards and string lines.

The first stage of the process is the removal of "oversite"[①]. Following the removal the oversite, excavation of the foundation trenches will take place to the required depth. On very small sites this may be undertaken by hand; however, the use of compact mechanical diggers makes mechanical excavation practical and economical on even modest sites. For larger sites it is the norm to use mechanised plant for excavation.

The excavation of foundation trenches follows an established sequence of operations,

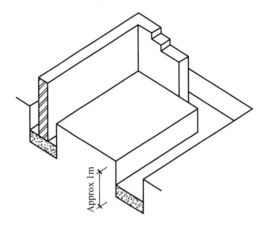

Fig. 9 - 3　Traditional strip footing foundation

which ensures that they are formed to an appropriate standard and with required levels of safety for the building operatives.

As the loading exerted on the foundations increases, or the load-bearing capacity of the ground decreases, so the required foundation width, to avoid overstressing of the ground, increases. In order to resist shear failure inside foundations, they would be required to be very thick, for the reason previously described; this would normally be uneconomic. Hence, to reduce the required foundation thickness while still resisting failure in shear or bending, steel reinforcement is introduced to the regions subjected to tension and shear. This provides a cheaper and more appropriate solution. Such a formation is termed a wide strip foundation (Fig. 9 - 4).

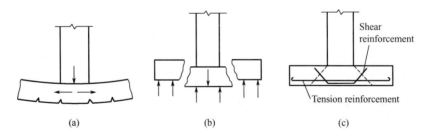

Fig. 9 - 4　Design of wide strip foundations to avoid failure in shear and bending
(a) Foundation fails through bending; (b) Foundation fails through shear;
(c) Reinforced foundation resists failure

2. Pad foundations

In some instances, such as when constructing framed buildings, the loads from the building are exerted upon the foundations in the form of concentrated point loads. In such cases, independent pad foundations (Fig. 9 - 5) are normally used. The principle of pad foundation design is the same as that of strip foundations, i. e. the pad area will be of sufficient size to safely transfer loadings from the column to the ground without exceeding its load-bearing capacity. The pad will generally be square, with the column loading exerted centrally to avoid rotation, and is normally formed from reinforced concrete; although other formations such as steel are possible, they are rare in modern construction. As a result of the concentration of loading applied to pad foundations, it is vital that they are reinforced sufficiently against shear as well as bending.

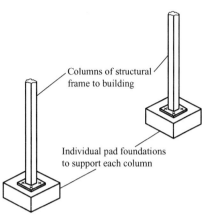

Fig. 9-5 Pad foundations

3. Raft foundations

Raft foundations (Fig. 9-6) are constructed in the form of a continuous slab, extending beneath the whole building, and formed in reinforced concrete. Hence the loadings from the building are spread over a large area and avoid over stressing the soil. The foundation acts, quite literally, like a raft. Inevitably, however, there tends to be a concentration of loading at the perimeter of the slab, where external walls are located, and at intermediate points across the main surface, resulting from loadings from internal walls etc. Hence the raft is thickened at these locations and/or is provided with extra reinforcement. At the perimeter (Fig. 9-7), this thickening is termed the "toe" and fulfils an important secondary function in preventing erosion of the supporting soil at the raft edges. If not prevented, undermining of the raft could occur, with serious consequences.

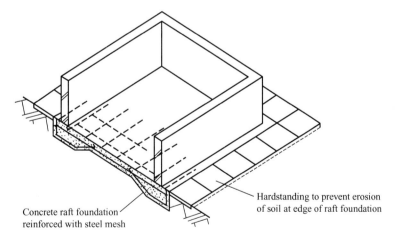

Fig. 9-6 Raft foundations

The edge detailing of raft foundations is dictated by the nature of the loading applied. Here we see four alternative options (Fig. 9-7) that are in common use — these show different structural approaches and details of exclusion of moisture are omitted for clarity. It must also be noted that edge protection is required where the soil tends to potential corrosion; this is often in the form of a concrete apron or concrete slab extending from the

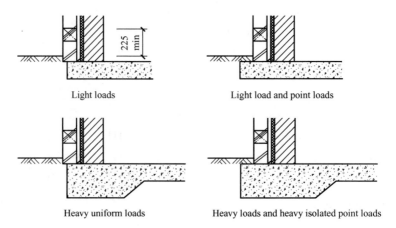

Fig. 9-7 Raft edge details

edge or "toe" of the raft.

Foundations of this type are generally utilized in the construction of relatively lightweight buildings, such as houses, on land with low bearing capacity. One of the advantages of this form of foundation is that if settlement occurs, the building moves as a whole unit, differential movement is prevented, hence, the building retains its integrity.

Ⅰ. New Words

1. stratum *n.* (pl. strata) 地层，岩层
2. trench *n.* 沟，沟槽

Ⅱ. Phrases and Expressions

1. point load   集中荷载
2. uniformly distributed load   均布荷载
3. load-bearing wall   承重墙
4. strip foundation   条形基础
5. pad foundation   独立基础

Ⅲ. Notes

① "oversite" 一词，这里用来指包含植被层的表层土壤。这层土必须被移除，以确保建筑物是建在稳定的、不会随时间的推移而易受改变的地层上。

Ⅳ. Exercises

1. Explain how shear failure can occur in a strip footing foundation and how this dictates certain dimensions of the concrete.

2. The sequence of operations involved in excavation for the placement of foundations is important. From the following list of activities select the operations involved in the appropriate order.

_____ Excavation of trench.

_____ Placement of concrete.

_____ Levelling and ramming trench bottoms.

_____ Setting out.

_____ Excavation of oversite.

_____ Timbering of trench sides.

_____ Trimming of trench sides.

3. Why is a deep strip footing only suited to certain types of soil?

4. Translate the expressions into English.

(1) 条形基础　(2) 独立基础　(3) 筏形基础　(4) 均布荷载　(5) 集中荷载

Ⅴ. Expanding

Know about the following terms related to pile foundation.

1. friction pile   摩擦桩
2. end-bearing pile   端承桩
3. displacement pile   打入桩，挤土桩
4. solid pile   实心桩
5. precast pile   预制桩
6. cast-in-place pile/cast-in-situ pile   灌注桩，现浇桩
7. load-supporting pile   承载桩
8. continuous flight auger   连续旋翼式螺钻

Ⅵ. Reading Material

Deep Foundations Forms

1. Introduction

After studying this section you should have developed an understanding of the nature and functions of deep foundations to low-rise construction. You should understand the nature and construction form of deep foundations. In addition, you should be able to identify and evaluate the factors that affect their choice.

2. Overview

The use of deep foundation forms for the construction of dwellings is relatively uncommon. The expense of deep foundation formation is one reason for this. However, the main reasons are functional, since the relatively low loads exerted by dwellings are normally dealt with effectively by shallow forms, such as strip foundations. The increasing need to develop sites with less favorable ground conditions has resulted in an increased use of deeper foundation forms. These are needed to cope with sites where ground quality close to the surface is poor or variable. Hence in some circumstances deeper foundation forms are essential.

3. Piles

Piles are often described as columns within the ground, since the basis upon which they work is similar to that of a traditional column, in that they transfer loadings from a higher level to a load-bearing medium at a lower level. They can be categorized in two ways (Fig. 9 – 8): by the way in which they are installed, or by the way in which they transfer their loads to the ground. Hence the following classifications are used to define pile types.

Definition by installation method: displacement piles (driven), replacement piles (bored).

Definition by load transfer mechanism: friction piles, end-bearing piles.

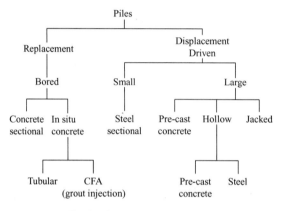

Fig. 9 – 8 Range of pile types

1) Displacement piles

Displacement piles are set into the ground by forcing or driving a solid pile or a hollow casing to the required level below ground, thus displacing the surrounding earth. In the case of solid piles, precast piles of required length may be driven into the ground using a driving rig, or alternatively the rig may be used to drive a series of short precast sections, which are connected as the work proceeds. The use of the second of these methods is by far the most efficient, since the length which is required may vary from pile to pile. Thus the use of one-piece precast piles inevitably necessitates adjustment of length when extending of one-piece piles is a difficult task on-site.

The difficulties of providing piles of exactly the correct length, together with the danger of damage to the pile resulting from the percussive driving force, have resulted in the adoption of the use of driven shell, or casing, piles. With this method (Fig. 9 – 9), a hollow shell or casing is driven into the ground, using a percussive rig, in a number of short sections; concrete is then poured into the void as the casing is withdrawn, the steel reinforcement having already been lowered into the hole. By this method, an in situ pile is formed in the ground. The vibration of the pile casing as it is withdrawn from the ground, results in the creation of a ridged surface to the pile sides, thus taking the most advantage of any frictional support provided by the ground.

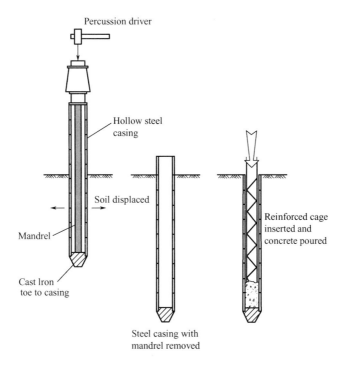

Fig. 9-9 Installation of displacement piles

Whether in the form of a solid pile or a hollow casing, the driving of the pile/casing is aided by the use of a driving toe or shoe, often in the form of a pointed cast-iron fitting at the base of the pile to allow easier penetration of the ground. These methods of installation have several disadvantages, in that considerable levels of noise and vibration are generated as a result of the driving operation. Hence they are generally considered unsuitable for congested sites, where adjacent buildings may be structurally affected, or areas where noise nuisance is undesirable. However, they may be used to good effect in consolidation of poor ground, by compressing the earth around the piles.

2) Replacement piles

Unlike displacement piles, replacement piles are installed by removing a volume of soil and replacing it with a load-supporting pile (Fig. 9-10). The holes are bored either by using a hollow weighted grab, which is repeatedly dropped and raised, removing soil as it does so, or by using a rotary borer or auger. As the excavation progresses, the sides are prevented from collapsing by introducing a shell, or casing, normally made from steel, or by the use of a viscous liquid called bentonite. The bentonite is then displaced by concrete as it is poured into the excavation and is stored for further use. This method is quieter than the displacement method and does not result in damage to surrounding buildings.

3) Friction piles and end-bearing piles

The method by which the pile transfers its load to the ground depends upon the nature of the design of the pile and is dictated to a large extent by the nature of the ground in which it is

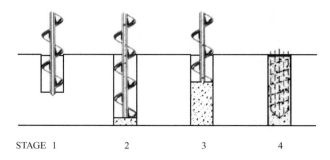

STAGE 1　　　　2　　　　3　　　　4

Fig. 9 - 10　Installation of replacement piles

located. Piles are used as a form of foundation in a variety of situations, each imposing differing demands upon pile design and creating restrictions on the ways in which the piles act. Although there are a great variety of situations which may necessitate the use of piles, the following are some of the most common:

(1) Where insufficient load-bearing capacity is offered by the soil at a shallow depth, but sufficient is available at a greater depth;

(2) Where the nature of the soil at a shallow depth is variable and performance is unpredictable, such as in areas of filled land;

(3) Where soils at shallow depths are subject to shrinkage or swelling due to seasonal changes;

(4) Where buildings or elements are subjected to an uplifting force, and require to be anchored to the ground.

It can be seen that in some of these instances one form of pile may be obviously more appropriate than another as a result of the way in which they act. End-bearing piles act by passing through unsuitable strata to bear directly upon soil with adequate bearing capacity, while friction piles are supported by the effects of friction from the ground to the sides of the pile throughout its length (Fig. 9 - 11); hence the formation of ridges to the sides of piles as described earlier. In practice, all piles derive their support from a combination of these factors.

The nature of operation of the end-bearing pile is such that it acts as column transferring load to stable strata beneath. The friction pile does not rely on the ability to sit on firm strata. Instead it relies on the friction from the ground surrounding it along its length. In practice both forms derive some load-bearing capacity from each of these mechanisms; friction and end bearing to a greater or lesser extent. A fundamental requirement is they are fully laterally restrained to avoid failure by buckling.

4) Connections to piles

The processes of installation of piles result in the tops of the piles being far from perfectly level and true; hence a loading platform must be created to take the loads from the building and transmit them safely to the piles. This platform is known as a pile cap,

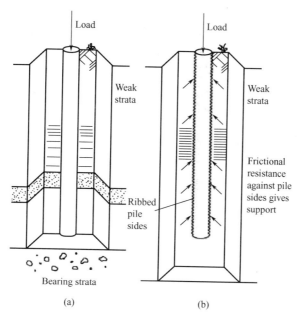

Fig. 9 – 11 End-bearing and friction piles
(a) End bearing pile; (b) Friction pile

which is formed in reinforced concrete and may transfer loads to a single pile or a group of piles. The caps are normally loaded by the columns of the building and may also take loads from the walls via a ground beam (Fig. 9 – 12).

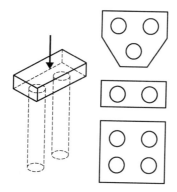

Fig. 9 – 12 Pile cap configurations

The ground beam is set into the ground to form a base for the construction of the walls and the load is transferred from the ground beam into the piles. The size of the beam and the size, depth and spacing of the piles are dictated by the nature of the ground and the extent of the loading from the structure. Structural calculations will be carried out for each individual case to arrive at a bespoke solution.

The ground beam spans between pile supports. In some soils there is risk of swelling which could cause heave beneath the beam. This would create problems and is avoided by the placement of a compressible layer between the beam and the soil.

参 考 译 文

第9课 浅基础形式

基础的性质使其能够适用于某种特定情况，这取决于诸多因素。荷载的大小及其作用在基础上的方式和持力层的性质等决定了基础设计。

一般而言，适用于住宅建筑的基础方案仅限于简单的形式；通常定义为浅基础和深基础两大类。这里我们将讨论浅基础。

浅基础有许多不同的形式，选择时应考虑所选结构形式的性质、地基可能承受的荷载及其承载力。

1. 条形基础

最常见的基础形式就是简单的大体积混凝土条形基础。当条形基础用于要建造的连续墙下时，其应包括一条足够宽的大体积混凝土带，将作用在其上的荷载扩展到与场地的接触面上，以确保不会出现超限应力。在条形混凝土基础内产生的压应力，其分布趋于在墙脚边起沿45°线延伸而围成的区域内。因此，为了避免沿45°线的冲切破坏，基础高度必须是这样的，即基础底部的有效宽度应落于这一区域内（图9-1）。

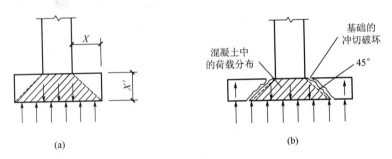

图9-1 避免冲切破坏的基础设计
(a) $X' \geqslant X$ 以避免冲切破坏；(b) 持力层的反作用

施加于基础中心区域的荷载分布于45°角范围内的混凝土内［图9-1(a)］。这样就能够在基底得到一个特定的受荷区域。荷载由地基反力来抵抗。如果基础太宽而导致地基反力作用超出了这一区域以外，就可能发生冲切破坏。为了避免冲切破坏［图9-1(b)］，在设计基础时应确保基础不会延伸至这一向下受荷区域以外。基础的宽度取决于地基土的强度与施加荷载的大小（即调整基底面积以减小压应力）。随着基础宽度的增加，基础高度也应增加，以避免冲切破坏。

条形基础主要用于建筑物传来的荷载相对适中且分布均匀的建筑，如住宅建筑（图9-2）。由于住宅的结构荷载通过外墙传递到基础上，所以它们一般是均匀分布的。在某些建筑形式中，荷载是以局部或集中荷载的形式传递的。在这种情况下，可采用独立基础来承受集中荷载，这在住宅建筑物中很少见。

大多数的传统房屋建筑都会使用条形浅基础。基础的宽度取决于地基土的安全承载力及受荷的大小。基础的宽度应调整至可提供足够的接触面积从而防止土体出现超限应力。

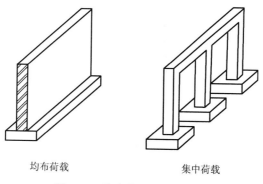

均布荷载　　　　　集中荷载

图 9-2　均布荷载与集中荷载

基础埋深与持力层的深度有关，通常应超过 1m（图 9-3），以避免接近地面的土层因冻结而隆起或收缩。

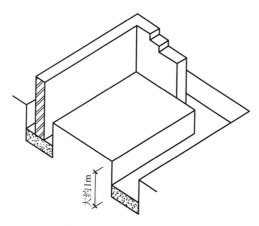

图 9-3　传统的条形基础

当建筑物位置测定后，用轮廓板和细线（轴线白线）确定沟槽位置，进行基础位置的开挖。

此过程的第一阶段是移除表层部分。随着表层部分的移除，基础沟槽将被挖至所需深度。在较小的场地，可以采用人工开挖；但是，即使在大小适中的场地，使用小型挖掘机进行机械开挖还是经济实用的。对于大型的场地，使用机械化设备进行开挖则是标准的做法。

基础沟槽的开挖应遵循已制定的操作程序，以确保依据适当的标准并符合建筑人员所要求的安全标准开挖基槽。

随着作用于基础上的荷载的增大，或者地基土承载能力的降低，基础宽度也应该增大，以避免地基土中的应力超限。由于上述原因，为防止扩展基础出现剪切破坏，基础必须很厚，而这通常是不经济的。因此，为了防止剪切破坏或弯曲破坏，要减小基础所需厚度，应在受拉区和剪切区配置钢筋。这是一个更经济适用的解决方法。这种形式被称为条形扩展基础（图 9-4）。

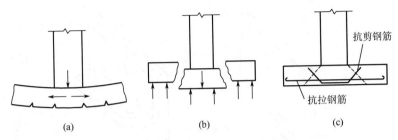

图 9-4 条形扩展基础的抗弯与抗剪设计
（a）基础的弯曲破坏；（b）基础的剪切破坏；（c）加筋基础防止破坏

2. 独立基础

在某些情况下，比如建造框架结构建筑时，建筑物的荷载以集中荷载的形式作用在基础上。在这种情况下，通常采用独立基础（图 9-5）。独立基础的设计原则与条形基础一样，即基础的尺寸满足将荷载从柱子安全地传至地基且不超出其承载能力。独立基础通常是正方形的，柱荷载应作用在其形心位置以避免倾覆，并且通常由钢筋混凝土构成；尽管其他的形式如钢结构也有用到，但在现代建筑中却很少见。因为独立基础承受集中荷载，所以基础配置钢筋来有效抵抗剪切和弯曲破坏至关重要。

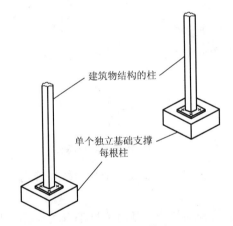

图 9-5 独立基础

3. 筏板基础

筏板基础（图 9-6）是指用钢筋混凝土建造的跨越整个建筑下方的一块连续板。因此，建筑物的荷载遍布很大一个区域且不会在地基土中产生过大的应力。从字面上看，这种基础的形状像一个木筏。但是，不可避免地，在外墙所在的平板周边会出现集中荷载，而且由于内墙等传来的荷载，在筏板表面的中间部位也会出现集中荷载。因此，在这些部位应该增厚筏板基础或者增设附加钢筋。在周边（图 9-7），这种加厚措施被称为"坡脚"，它能够实现一个重要的辅助功能，即防止筏板边缘被土壤腐蚀。如果无法阻止，筏板会被掏空，这会导致非常严重的后果。

筏板基础的边缘细部构造是由荷载性质决定的。图 9-7 为四种常用的选择方案；这些方案显示的是不同的构造做法，为清晰起见而忽略了排水细节。必须注意的是，在易遭

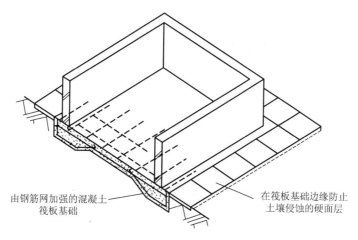

图 9-6 筏板基础

受土壤潜在侵蚀的部位，筏板边缘应进行必要的防护；通常是采用混凝土护壁或者是从筏板边缘或"坡脚"延伸出的混凝土板。

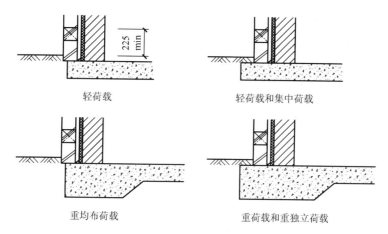

图 9-7 筏板边缘细部构造

此类基础通常用于自重较小而地基承载力较低的建筑物，如住宅建筑。这类基础的优点之一是如果发生沉降，建筑物会整体沉降而不会发生不均匀沉降，因此能够保持建筑物的整体性。

Lesson 10

Structural Concepts

The building fabric, having been broadly conceived in terms of an environmental envelope, must be of such a nature that it can withstand safely all the forces to which the building will be subjected in use. In other words it must be developed as a structure, a fabrication which for practical purposes does not move in any appreciable manner under its loads. Buildings vary widely in form and appearance but throughout history they have all developed from three basic concepts of structure. These are known as skeletal, solid and surface structures.

1. Skeletal structure

As the term implies this consists essentially of a skeleton or framework which supports all the loads and resists all the forces acting on the building and through which all loads are transferred to the soil on which the building rests. Simple examples are the North American Indian and the mid-European wigwams in which a framework of poles or branches supports a skin or tree bark enclosing membrane (Fig. 10 – 1). This elementary form has developed throughout history into frameworks which consist essentially of pairs of uprights supporting some form of spanning member as shown in Fig. 10 – 1. These are spaced apart and tied together by longitudinal members to form the volume of the building. In these frames the vertical supports are in compression. <u>Skeletal structures in which the floors are suspended from the top of the building by vertical supports in tension are generally called suspension structures.</u>① Other forms of the skeletal structure are the frameworks or lattices of interconnected members known as grid structures, an example of which is shown in Fig. 10 – 1.

By its nature the skeleton frame cannot enclose the space within it as an environmental envelope and other, enclosing, elements must be associated with it. The significance of this clear distinction between the supporting element and the enclosing element is that the latter can be made relatively light and thin and is not fixed in its position relative to the skeleton frame—it may be placed outside or inside the frame or may fit into the panels of the frame as may be seen in examples of contemporary steel or concrete frame structures. The practical implications of this distinction are discussed later. Skeletal structures are suitable for high-and low-rise, and for long-and short-span buildings.

2. Solid structure

In this form of structure the wall acts as both the enclosing and supporting element. It

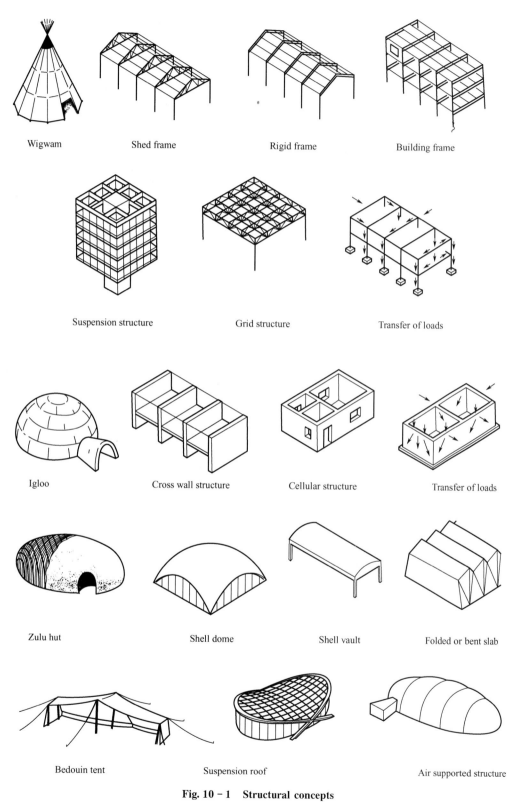

Fig. 10 − 1 **Structural concepts**

falls, therefore, within the category of load bearing wall structures, an inclusive term implying a structure in which all loads are transferred to the soil through the walls. The characteristic of this particular form is a wall of substantial thickness due to the nature of the walling materials and the manner in which they are used, such as in masonry and mass concrete work.② The Eskimo igloo is an interesting example of this type of construction (Fig. 10 - 1) although for technical and economic reasons circular plan forms have been less used than rectangular forms for buildings constructed in this way. Solid construction in the form of brick and stone wall buildings has been used over the centuries and, in certain circumstances, in its various modem forms it is still a valid and economic type of construction for both high-and low-rise buildings if these are of limited span permitting types of floor structure which impose an even distribution of loading on the wall (Fig. 10 - 1).

Roof structures are not vaulted over in solid construction, even over limited spans such as that of the Eskimo igloo, because of the problems of construction and the existence of cheaper, lighter and more quickly erected alternatives.

3. Surface structure

Surface structures fall into two broad groups: (1) those in which the elements are made of thin plates of solid material which are given necessary stiffness by being curved or bent, and (2) those in which the elements consist of very thin flexible sheet membranes suspended or stretched in tension over supporting members. A Zulu woven branch and mud hut (Fig. 10 - 1) and modern reinforced concrete shell and folded slab structures are typical of the first. In this form also the wall, and the roof, may act as both the enclosing and supporting structure but the manner in which particular materials are used results in quite thin wall and roof elements.③ Those in the second group are used for roofs and are known as tension structures. One form is typified by the traditional Bedouin tent (Fig. 10 - 1) of which delightful modern applications were first made by Frei Otto for roofing temporary exhibition buildings. Utilizing suitably developed membranes this form can now be used for roofing permanent structures. Another form in this group, using compressed air as the supporting medium for similar types of membrane, dispenses with compression members over which, in the tent form, the membrane is stretched. In this the membrane is fixed and sealed at ground level and is tensioned into shape and supported by air pumped into the interior and maintained under slight pressure (Fig. 10 - 1). Alternatively, inflated tubes may be incorporated which form supporting ribs to the membrane stretched between them. These are called air-stabilized or pneumatic structures. In a third form in this group the membrane consists of steel cables suspended from supports and carrying a thin applied cladding and weatherproof covering (Fig. 10 - 1).

Surface roof structures are particularly economic over wide spans and for a more detailed consideration of their reference should be made in latter study.

Lesson 10 Structural Concepts

Ⅰ. New Words

1. fabric *n.* 构造，建筑物，织物，组织
2. skeletal *adj.* 骨骼的，骨架的，框架的
3. framework *n.* 构架，框架，骨架，结构
4. transfer *vt.* 搬，转移，传递
5. longitudinal *adj.* 长度的，纵向的，经度的
6. compression *n.* 压缩，压迫，浓缩
7. lattice *n.* 格子，格栅，格架
8. contemporary *adj.* 同时代的，现代的，同一时期的
9. span *n.* 跨度，跨距
10. exert *vt.* 运用，发挥，施加
11. stretch *vt.* 伸展，张开，展开
12. inflated *adj.* 膨胀的，充气的

Ⅱ. Phrases and Expressions

1. building fabric   建筑结构
2. skeletal structure   骨架结构
3. grid structure   网架结构
4. vertical support   竖向支撑
5. tension structure   张拉结构
6. suspension structures   悬索结构
7. solid structure   实体结构
8. in certain circumstance   在某种情况下
9. high-rise building   高层建筑
10. low-rise building   低层建筑
11. pneumatic structure   充气结构

Ⅲ. Notes

① "are suspended from" 译为 "悬挂于……，被悬浮在……"。"suspension" 是 suspend 的名词形式，译为 "悬挂、悬吊、悬浮"。

② "due to" 本意为 "归功于……，归结于……"，在这里译为 "由……决定"。该句可理解为：这种特殊形式（实体结构）的特点是墙体的实际厚度由墙体材料的性质及所用的形式决定，如砌体结构和大体积混凝土工程。

③ 此句中 "in which" 引导的定语从句作修饰成分，但是由于专业英语的句式较长，意义复杂，在翻译的过程中常常把定语从句与主句分开翻译，故此句可译为：在这种结构形式中，墙体和屋顶可充当围护和支承结构，但由于特定材料的应用形式不同导致墙体和屋顶构件都非常薄。

Ⅳ. Exercises

Please write down the name of each structural form shown in Fig. 10 – 2.

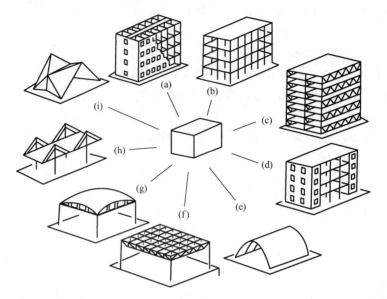

Fig. 10 – 2 The forms of structure

Ⅴ. Expanding

Know about the other terms related to the structural forms.

1. bearing-wall structure   承重墙结构
2. frame structure   框架结构
3. shear wall structure   剪力墙结构
4. frame-shear wall structure   框架-剪力墙结构
5. truss structure   桁架结构
6. tube structure   筒体结构
7. arch structure   拱结构
8. bent frame structure   排架结构
9. folded plate structure   折板结构
10. membrane structure   薄膜结构

Ⅵ. Reading Material

Structural Engineering

Structural engineering is a branch of civil engineering concerned with the designing and execution of all types of structures, such as buildings, bridges, highways, power plants, dams, transmission towers, and many other kinds of specific structures.

The designing phrase starts with the understand of the project. The designer must take a thorough study of the technological and service performance requirements that must be expected from the structure, including load intensities and their duration, any dynamic action that might take place.

Load condition is the first factor that designers consider. The same structure in different location exhibits different design because of ground water level, soil characteristic. Foundation is particularly important in whole structure design. If the soil is soft, it should be strengthened. If the substructure is below the grounder water level, methods such as well points, or pumping from sumps should be taken to remove water.

A designer must first calculate the dead loads, live loads, earthquake and wind loads, and their combination, then selects structural system and construction materials. Finally, the designer analyzes structure and designs members. The live loads are usually provided by building codes. Steel and concrete are traditional materials; carbon fiber is novel material which has excellent strength and stiffness, but they are used only in limited application because of the high cost.

Analysis of structures aimed at determining the forces and deformations existed in members. These forces such as tension, compression, bending, shear and torsion could make structures destroyed. Excessive lateral sway may cause recurring damage to partitions, ceilings, and other architectural details and may cause discomfort to the occupants of the building. This deformation must be kept within acceptable limits.

Structural design and structural analysis are components of structural engineering. They are interlocked subjects. The structural engineering has the objective of proportioning a structure so that it can safely carry the loads to which it may be subjected. Structural analysis provides the internal forces and structural design utilizes those forces to proportion the members or systems of members. Without structural analysis design is impossible.

Member sizes designed are often from experience and comparison to some similar design and use of available empirical rules combined with some rough calculations. Most design are initially based on the strength and stability criteria, while other criteria are used to carry out checks at a later stage. To arrive at an optimum and economical design, it is usually to repeat the analysis with the revised sizes and shapes. In this stage, computer is a useful tool.

Ways of constructing the building fabric, that is to say the manner in which it is formed of different materials, vary with (1) the structural concept on which it is based, (2) the nature of the materials used and (3) the manner in which the materials are combined. For example, if solid load-bearing wall construction is adopted this may be constructed of masonry units or of concrete: the type of masonry units can vary and may be combined in different ways and the concrete may be formed into walls by in situ casting or by precasting. The form of construction will also vary with the functional requirements of particular parts of the fabric since these may be satisfied by various materials in varying combinations. As explained before, for example, adequate weather resistance can be provided in external walls by using either solid masonry of considerable thickness or by using a thinner wall incorporating a cavity which prevents the passage of moisture from the external, wet face to the interior face of the wall.

Different forms of construction are, fundamentally, organizational devices used for economic reasons. They vary with the availability and the relative costs of building resources, especially of labour and materials and develop for reasons of economy of time and labour and of materials. Over the course of history building materials have been, and very largely still are, heavy and bulky and the earliest buildings were constructed of local materials in the absence of cheap and easy means of transport. As the supply of these materials became locally depleted and the need to import from other areas arose the economic use of materials became increasingly important and new forms of construction were introduced, developing from a better understanding of these materials, in which they were used more economically, thus requiring less labour in obtaining them and less transport.

Economy of labour in actual building also exercises considerable influence on forms of construction, either because of rising costs of labour or scarcity of labour. Thus, as building passed from the "self-build" stage of the early building days into the "contractor-built" era, involving paid labour, forms of construction developed in which less labour was required, for example, the development of brick construction to supersede labour-intensive forms such as traditional rammed earth construction for walling.

As well as the actual cost of labour the relative costs of labour and materials can have a significant effect upon forms of construction. Where the cost of labour is considerably higher than that of materials methods tend to develop in which the labour content of the building operations is reduced at the expense of an increase in the amount of material used, as in the American "plank and beam" form of timber floor and roof construction which uses large, widely spaced joists or beams spanned by thick boarding involving a greater timber content than forms of floor construction but involving less labour in fabrication.

Scarcity of labour, particularly skilled labour, has a similar effect to that of rising costs of labour. Both bring about forms of construction which are economic of labour, such as the use of concrete blocks instead of bricks for masonry work because these are quicker to lay or the use of modern trussed rafters for small span timber roof construction which greatly reduces the labour content of site fabrication compared with that of traditional methods. Because of the scarcity of plasterers in the past the use of plasterboard in place of lath and plaster has become an accepted technique.

Forms of construction which reduce labour content often decrease the time required for the operations, that is they result in increased productivity. When scarcity of labour and the need to increase productivity are current problems of the building industry the development of new forms of construction requiring less labour is an important exercise. But production can also be increased by good organization as well as by changes in construction and this has led to a general re-appraisal of the whole process of production in the field of building.

参考译文

第 10 课　结构概念

建筑结构，从被广泛理解的包围环境的角度来看，必须具有能够安全地承受使用期间施加于建筑物上的所有荷载的性质。换句话说，建筑结构必须被开发为一种结构，即在实际使用过程中不会在负荷之下发生任何明显位移的构造。建筑物的形式和外观各不相同，但是，纵观历史，所有建筑物均是基于三种基本结构概念发展而来的。这三种基本结构概念包括骨架、实体及表层结构。

1. 骨架结构

顾名思义，骨架结构本质上是由支承所有负荷并抵抗施加于建筑物上的所有（作用）力的骨架或框架组成，并且所有负荷均通过骨架结构被传递至建筑物下的土层。一些骨架结构的简单例子有北美印第安人和中部欧洲人修建的棚屋，他们使用柱框架或树枝支承由兽皮或树皮充当的围护膜结构（图 10-1）。这种基本结构形式在历史进程中已发展为基本由双立柱支撑某种跨越构件构成的框架结构，如图 10-1 所示。立柱被间隔开并通过纵向构件结合在一起，形成建筑物的整体。在此类框架结构中，竖向支承构件是受压的。楼板被受拉的竖向支承构件悬挂在建筑物顶部的骨架结构通常被称为悬吊结构。骨架结构的其他形式包括由互相连接的构件组成的框架结构或网格结构，被称为网架结构，示例如图 10-1 所示。

就其性质而言，骨架结构无法作为外围护结构封闭内部空间，必须使用其他围护构件。对承重构件与围护构件进行明确区分的意义在于，后者相对轻薄并且其相对于骨架结构的位置不固定——它可被置于框架外部或内部或被镶嵌在框架平面内，如可能在现代钢或混凝土框架结构中看到的例子。这种区别的实践含义将在下文中得到讨论。骨架结构适用于高层和低层建筑以及大跨度和短跨度建筑物。

2. 实体结构

在这种结构形式中，墙体同时充当围护和支承构件。因此，这种结构属于承重墙结构类型，这一广义性术语是指全部负荷均通过墙体传递至土层的结构。这种特殊形式（实体结构）的特点是墙体的实际厚度由墙体材料的性质及所用形式决定，如砌体结构和大体积混凝土工程。因纽特人冰屋属于此种建筑类型的一个有趣的示例（图 10-1），尽管由于技术和经济方面的原因，使用圆形平面形式的建筑已经比使用矩形平面形式的建筑要少。砖石墙体建筑形式的实体结构已经在多个世纪中得到应用，在某些情况下，在其各种现代结构形式中，如果这些允许跨度限值内的楼板结构给墙体施加均匀分布的荷载，那么此种结构仍然是一种可用于高层和低层建筑的有效而经济的结构类型（图 10-1）。

由于施工方面的问题以及存在更便宜、更轻、修建速度更快的结构可选择，所以在实体结构中屋顶结构不是拱顶结构，即使是超过极限跨度的屋顶结构（如因纽特人冰屋）亦是如此。

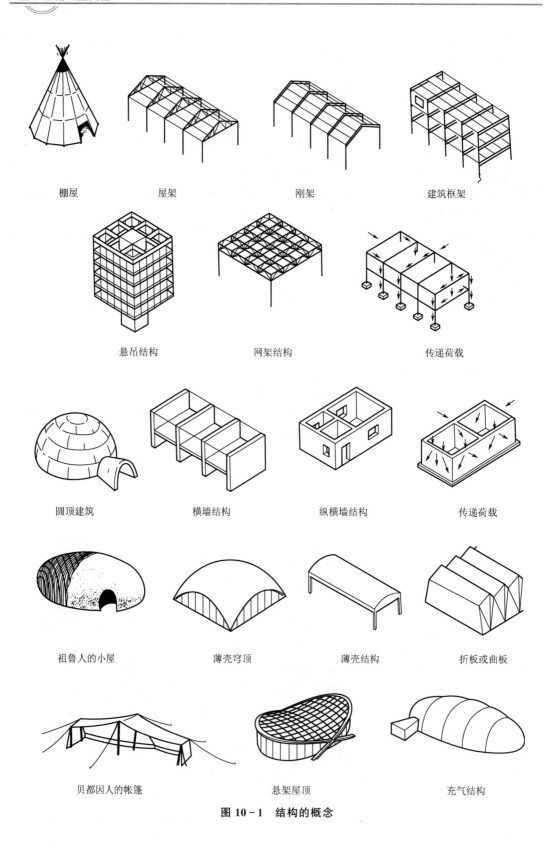

图 10-1 结构的概念

3. 表层结构

表层结构可被分为两大类：(1) 表层结构中的构件是由固体材料组成的薄板并通过弯曲的手段赋予材料必要的刚度，以及 (2) 表层结构中的各构件是在支承构件上悬挂或拉伸的非常薄的柔性片膜。祖鲁人使用编制树枝和泥修建的小泥屋（图 10-1）和现代钢筋混凝土壳体以及折板结构属于典型的第一类表层结构。在这种结构形式中，墙体和屋顶可充当围护和支承结构，但由于特定材料的应用形式不同导致墙体和屋顶构件都非常薄。第二类表层结构被用作屋顶且被称为张拉结构。传统贝都因人的帐篷即是此类结构形式的典型（图 10-1），弗雷·奥托首先对这种结构形式进行了很好的现代应用，用作临时展览建筑的屋顶。如今，合理利用成熟的薄膜（技术），这种张拉结构可用作永久的屋顶结构。此种结构中的另一种形式是使用压缩空气作为类似膜结构形式的支承媒介，无须受压构件，在帐篷式结构中，薄膜处于受拉状态。在这种结构中，薄膜在地面上经过固定和密封并被拉伸成型，由泵送入内部的空气提供支承并在轻微压力之下维持形状（图 10-1）。或者，可在该结构中使用充气管作为支承肋，为在其间拉伸的膜结构提供支承。此类结构被称为气稳或充气结构。对于此结构中的第三种形式而言，膜结构是由悬挂在支承结构上的钢拉索构成的，并且带有较薄的实用型覆盖层和全天候型覆盖层（图 10-1）。

表层屋顶结构对于大跨度建筑尤其具有经济性，在日后的研究之中应对这种形式均引用进行更为详尽的考虑。

Lesson 11

The Forces on a Building and the Effects

The forces on a building may be considered as having an overall effect on the building or structure which tends to move it as a whole and local effects on its parts which tend to deform them but not move them out of position. The stability of the building involves the equilibrium of all the forces acting on the structure and the absence of excessive deformations.

1. Overall movement

Vertical downward forces caused by the dead weight of the building and its loads tend to force it down into the soil on which it rests or tend to force floor or roof structures down on their supports.① In the first case the soil must be sufficiently strong to exert an upward force or reaction equal to the weight of the building (Fig. 11 - 1) and in the latter the supports must exert similar forces equal to that which the floor exerts on them (Fig. 11 - 2). Vertical active forces may also be upward as in the upward suction caused by wind passing over a flat roof which tends to raise the roof and its structure (Fig. 11 - 1).

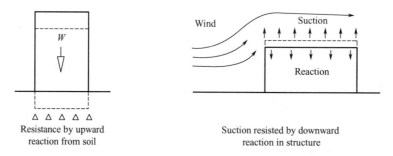

Fig. 11 - 1 Movement: vertical

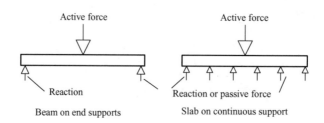

Fig. 11 - 2 Forces and reactions

Horizontal forces, which may be exerted by wind or soil against the side of a wall or a building, tend to make it slide on its base or overturn.② The former tendency must be resisted by the friction between the base and the soil on which the structure rests or by the passive pressure of the soil on the opposite side, the latter by the weight of the structure itself, by a strut or by a suitable tension element, any of which would cause a counter-moment (Fig. 11 – 3).

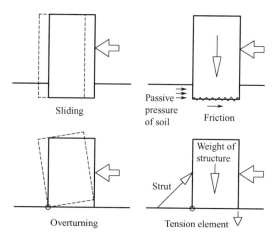

Fig. 11 – 3 **Movement: sliding and overturning**

Oblique forces have an effect similar to that of horizontal forces. Fig. 11 – 4 illustrates some circumstances in which forces are generated and their effects. A curved roof structure springing from its foundations will exert an oblique outward trust and the foundations will tend to slide outwards; the oblique trusts from an arch on fairly tall for supports will tend to overturn the supports. The oblique force from an inclined roof structure will tend to cause the roof both to slide off its supporting walls and to overturn the walls. The smaller the angle of inclination the greater is the tendency for sliding and overturning to occur since the line of the force more nearly approaches the horizontal. Methods similar in principle to those for horizontal forces are adopted to maintain the equilibrium of structures under oblique loading. In the examples shown a tension member, if used, would tie the two ends of the curved or inclined roof elements causing the oblique trusts and prevent them moving outwards. An oblique force acting downwards at any point generates both an outward horizontal force and a downward vertical force at that point, the magnitude of each of which will vary according to the inclination of the oblique force. These forces can be established graphically, with sufficient accuracy for practical purposes, by means of a parallelogram of forces as explained below.

2. Deformation

Vertical forces on thin walls or columns tend to make them bend in their height, in the same way as a thin stick under pressure on the top (Fig. 11 – 5). This is known as buckling. Similarly a load will cause vertical bending or deflection in a beam as in Fig. 11 – 5.

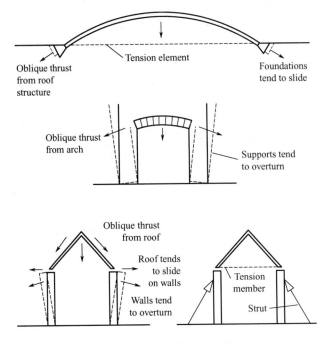

Fig. 11 – 4　Movement: effect of oblique forces

Buckling or sideways bending may also occur in the top of a thin, deep beam for reasons given later (Fig. 11 – 6).

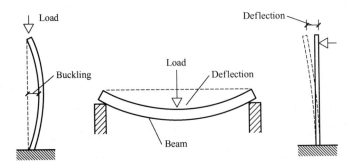

Fig. 11 – 5　Deformations

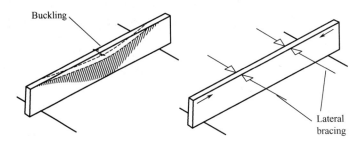

Fig. 11 – 6　Buckling in deep beams

Horizontal forces acting on thin walls or columns may cause deflection in them as they act like a vertical cantilever③ (Fig. 11-5). It should be noted that this is not identical to buckling which is caused by loading in the direction of its length, not normal to its length.

Buckling and deflection are both due to bending and both are controlled by using materials and members of adequate stiffness.

3. Triangle and parallelogram of forces

In the section on oblique forces reference is made to a method of establishing graphically the magnitude of forces acting on a structure. This is possible because a force may be represented in direction and magnitude by a line drawn parallel to the direction of the force and to scale so that its length represents the magnitude of the force.

When two forces act on a body, other than being directly opposed, they tend to move the body. Thus the joint effect of the two forces A and B in Fig. 11-7(1), acting in the direction shown, will be to move the body along a line somewhere between them, say along the broken line. A third force C, of an appropriate magnitude, acting along this line could have the same effect as the combined forces A and B and could replace them. Such a force is known as the resultant of the other two. A force D of exactly the same magnitude as C and directly opposed to it would produce equilibrium in the body and is known as the equilibrant of C and, consequently, of the two forces A and B.

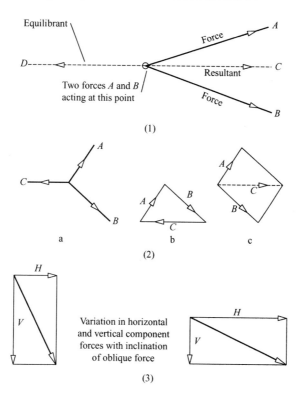

Fig. 11-7 Triangle and parallelogram of forces

In Fig. 11 - 7(2) a, the three forces are in equilibrium, therefore each is the equilibrant of the other two. If three such forces are drawn as lines to scale representing their magnitudes and parallel to their directions, continuing from each other in a clockwise manner as in Fig. 11 - 7(2) b, they will form a closed triangle. Thus if the direction and magnitude of only two, say A and B, are known the magnitude and direction of their equilibrant can be found as the closing side of the triangle. Such a triangle is known as a triangle of forces.

The resultant of any pair of the forces is, of course, equal but opposite in direction to their equilibrant and can often be found more conveniently by drawing a parallelogram of forces as in Fig. 11 - 7(2) c, where the forces A and B are drawn as the adjacent sides of a parallelogram, the diagonal of which will represent their resultant in magnitude and direction.

In practice it is often necessary not only to find the resultant of a pair of forces but to resolve a single force into two component forces as in the case of oblique forces. The known force is drawn to scale and in the correct direction as the diagonal of a parallelogram, the sides of which are drawn parallel to the directions of the required components. In Fig. 11 - 7(3) the diagonals represent oblique forces such as in Fig. 11 - 4 and the sides of the parallelograms represent the vertical and horizontal components. It will be clear from this that the smaller the angle of inclination of an oblique force, the greater will be the magnitude of its horizontal component and, therefore the greater the tendency for the structure to slide or overturn under its action.

Ⅰ. New Words

1. force *n.* 力，力量；*vt.* 促使，推动，强加
2. effect *n.* 结果，效应，作用，影响
3. deform *vt.* 使变形
4. equilibrium *n.* 平衡，均衡
5. oblique *adj.* 斜的，倾斜的
6. friction *n.* 摩擦，摩擦（力）
7. generate *vt.* 使形成，（使）产生，造成，引起，导致
8. slide *vi.* 滑，滑动，滑移
9. magnitude *n.* 大小，量级，震级
10. vary *vt.* 改变，使多样化
11. incline *vt.* 使倾斜，倾向于；*n.* 斜面，斜坡
12. graphically *adv.* 生动地，用图表表示，轮廓分明地
13. column *n.* [建筑学] 柱，柱状物，纵列
14. buckle *vt.* 弯曲；*n.* 屈曲，膨胀
15. bend *vt.* 使弯曲，屈服；*n.* 弯曲
16. cantilever *n.* 悬臂，悬壁；*vt.* 利用悬臂支撑
17. triangle *n.* 三角，三角形

Lesson 11 The Forces on a Building and the Effects

18. parallelogram *n.* 平行四边形

Ⅱ. Phrases and Expressions

1. excessive deformation   过量变形
2. horizontal force   水平力
3. passive pressure   受压
4. counter-moment   反力矩
5. oblique force   斜向力
6. vertical bending   垂直弯曲
7. vertical cantilever   垂直悬臂
8. be identical to   一致，与……相同
9. horizontal component   水平分量，水平分力

Ⅲ. Notes

① 该句中的"it"指代"building"；"its load"指代建筑物的负载。
② "which"引导一个非限定性定语从句，作为主句的补充说明。
③ "acting on thin walls or columns"在句子中作伴随状语，该句的谓语动词是"cause"。

Ⅳ. Exercises

Fill in the blanks with the information given in the text.

1. _____ on thin walls or columns tend to make them bend in their height, in the same way as a thin stick under pressure on the top.

2. _____ and _____ are both due to bending and both are controlled by using materials and members of adequate stiffness.

3. The smaller the angle of inclination of an oblique force, the _____ will be the magnitude of its horizontal component and, therefore the _____ the tendency for the structure to slide or overturn under its action.

Ⅴ. Expanding

Know about the following terms related to the limit state design method.

1. load effect   荷载效应
2. structural resistance   结构抗力
3. crack resistance   抗裂度
4. ultimate deformation   极限变形
5. the stiffness for a long time   长期刚度
6. short-term stiffness   短期刚度

Ⅶ. Reading Material

Load Effect and Structural Resistance

When a load Q is acting on the structure, it will produce load effect S, which is the response to the load in different parts of the structure, such as moment, shear stress strain, etc. Thus, it may be stated that S is a function of Q and may be expressed as

$$S=S(Q) \tag{1}$$

At the same time, to sustain that load effect, the structure has a capacity R that is inherent in the strength f of the material the structure is composed of. The capacity R depends also on the dimension s of the structure. Thus it may be stated that R is a function of s and f and may be expressed as

$$R=R(s,f) \tag{2}$$

Let it be defined that $Z=R-S$, then

When $Z>0$, $R>S$, structure reliable

when $Z<0$, $R<S$, structure failure

When $Z=0$, $R=S$, structure in limit state

Now, a structure is designed to be reliable, not to fail. But due to the fluctuation of the loads, the irregularities in the material strength, and the tolerance in the manufacture and the erection, none of the above factors is definite in their values. Besides, the mathematical functions of R and S may be empirical or approximate. So the design principle is to satisfy the relationship $R>S$ with some margin on the side of R. In the old practice, this was achieved by satisfying the equation $KS=R$, where $K>1$ and is called the factor of safety, and it is argued that this will make the structure "safe" from failure.

Now it is commonly realized that, theoretically, no structure can be designed to be absolutely reliable. What can be achieved is to design the structure so that the probability of its failure may be sufficiently low to be acceptable. Thus a structure destroyed in a violent earthquake, or destroyed by terrorist attack, although actually failed, can be accepted as "safe" if the chances of such contingencies are remote enough at the time of design. It is a subject for social psychological study as to what is an "acceptable risk". Whatever that value may be, the introduction of the probability concept and the reliability analysis in structural design is a significant step forward in design philosophy.

参 考 译 文

第 11 课 建筑物受力及其作用

作用于建筑物上的力可被视为能够对建筑物或结构产生全面作用,趋于使建筑物或结构发生整体运动,并且,这些力对建筑物或结构的各部件产生的局部作用易使此类部件变形,但不会使其离开所在位置。建筑物的稳定性包括确保施加于结构之上的所有力之间实现平衡且不存在过度变形。

1. 整体运动

由建筑物的自重及其负载所产生的垂直向下的力趋于向下施加给支承建筑物的土层,或趋于使楼板与屋顶结构向下施加给它们的支撑构件。在第一种情况下,土层必须足够坚实,以便向建筑物施加一个等于建筑物重量的向上的力或反作用力(图11-1),对于后者而言,支承构件必须能够施加与楼板对其施加的力相等的力(图11-2)。垂直作用力的方向也可以向上,比如刮过平屋顶的风导致的向上吸力,此种吸力趋于将屋顶及其结构向上拉起(图11-1)。

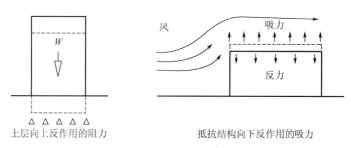

图 11-1 运动:垂直的

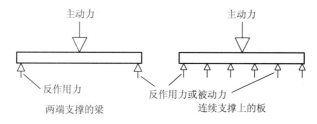

图 11-2 力与反作用力

由风或土层对墙体或建筑物侧面施加的水平力趋于使建筑物在地基上滑动或发生倾覆。前者(建筑物滑动)必须由基底与土层之间的摩擦力或施加于受力物体反面的被动土压力来抵抗,后者(建筑物发生倾覆)必须由能够产生反力矩的结构本身重量、支撑或适当的拉力构件来抵抗(图11-3)。

斜向力产生的作用与水平力产生的作用相似。图11-4表明了一些产生斜向力的情况及其对建筑物产生的作用。从基础架起的曲面屋顶结构会施加一个倾斜向外的力,会使基础趋于向外滑动,位于高处的支撑上的拱形结构产生的斜向力易导致支撑发生倾覆。坡屋面结构产生的斜向力可导致屋顶从支承墙上滑落或导致墙体发生倾覆。倾斜角度越小,发生滑动和倾覆的倾向越大,因为力的作用线更接近于水平方向。原则上,可以采用与水平力类似的方法来保持倾斜载荷下结构的平衡。示例中表明如果使用受拉杆件就能够固定导致斜向力的曲面或坡屋顶构件的两端,并防止两端向外运动。任何一点上的向下作用的斜向力均会在该点上产生一个向外的水平力和一个向下的垂直力,力的大小随斜向力倾角的不同而变化。此类力可建立图表表示,为了满足实践应用对于准确性的要求,可通过下文所述的力的平行四边形进行表述。

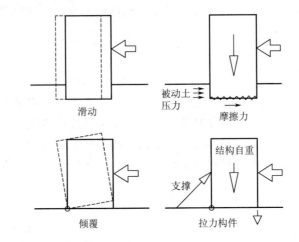

图 11-3 运动：滑动及倾覆

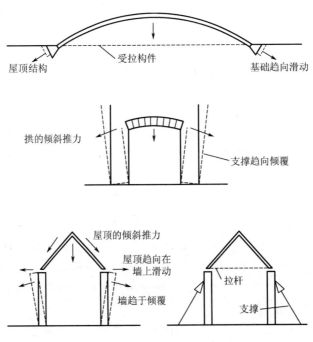

图 11-4 运动：斜向力的作用

2. 变形

施加于较薄的墙体或柱子上的垂直力可导致墙体或柱子在高度方向上发生弯曲，如同顶端处于压力之下的细杆发生弯曲一样（图11-5）。此种情形被称为屈曲。同样，负荷会导致横梁发生垂直弯曲或变形，如图11-5所示。宽度较小的深梁的顶部同样会由于下文所述的原因而发生屈曲或侧向弯曲（图11-6）。

施加于较薄的墙体或柱子上的水平力可导致墙体或柱子发生变形，因为此类墙体或柱子的受力表现如同垂直悬臂一样（图11-5）。应注意的是，此类变形与长度方向上的负荷导致的屈曲不同，并非垂直于长度。

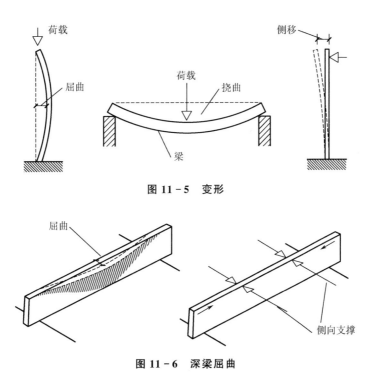

图 11-5　变形

图 11-6　深梁屈曲

屈曲和挠曲均是由于弯曲导致的，且均可通过使用具有足够刚度的材料和构件进行控制。

3. 力的三角形和平行四边形

在关于斜向力的部分中，可参考以图形建立作用在结构上的力的大小的方法。这是可能的，因为力可以通过平行于力的方向按比例绘制的线来表示方向和大小，所以，线的长度可以表示力的大小。

当两个力作用于一个物体时，除了两个力恰好相反的情况以外，二者趋于使物体发生运动。因此，图 11-7(1) 所示的两个力 A 和 B 在如图所示的方向上产生的共同作用将使物体沿二者之间的某条线发生运动，例如沿虚线方向。沿这条线作用的第三个大小适当的力 C 能够产生的作用与 A 和 B 的合力产生的作用相同并能够替代 A 和 B。此种力被称为上述两个力的合力。大小与力 C 相等并且恰好与其相反的一个力 D 能够使物体实现受力平衡，并被称为力 C 的平衡力以及力 A 和 B 的平衡力。

在图 11-7(2)a 中，三个力处于平衡状态下，因此，每个力均为其他两个力的平衡力。如果用线条按比例表示三个力［表明力的大小且与力的方向平行，按顺时针方向彼此连续，如图 11-7(2)b 所示］，则线条会形成一个封闭的三角形。因此，如果仅仅已知两个力（如 A 和 B）的方向和大小，则二者平衡力的大小和方向即是三角形闭合边。此类三角形被称为力的三角形。

当然，任何成对的力的合力等于其平衡力，但方向与其平衡力相反，且通常可以通过绘制如图 11-7(2)c 所示的力的平行四边形更方便地找到。其中，力 A 和 B 被绘成平行四边形的邻边，平行四边形的对角线代表二者合力的大小和方向。

通常，在实践中不仅需要找到一对力的合力，而且还需将一个力分解为两个分力，例

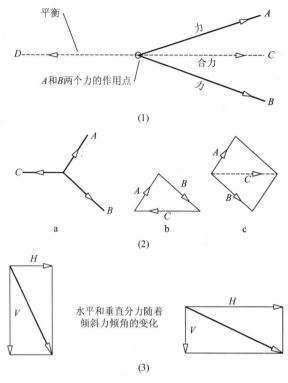

图 11-7 力的三角形和平行四边形

如在斜向力的情况下。按照比例和正确的方向将已知力绘成一个平行四边形的对角线,平行四边形的边与分力的方向平行。在图 11-7(3) 中,对角线代表斜向力(图 11-4),并且平行四边形的边代表垂直和水平分力。通过该图可知,斜向力的倾斜角越小,其水平分力越大。因此,该结构在力的作用下发生滑动或倾覆的趋势越大。

Lesson 12
Structural Design

Structural design is the selection of materials and member type, size, and configuration to carry loads in a safe and serviceable fashion. In general, structural design implies the engineering of stationary objects such as buildings and bridges, or objects that may be mobile but have a rigid shape such as ship hulls and aircraft frames. Devices with parts planned to move with relation to each other (linkages) are generally assigned to the area of mechanical design.

Structural design involved at least five distinct phases of work: project requirements, materials, structural scheme, analysis, and design. For unusual structures or materials a sixth phase, testing, should be included. These phases do not proceed in a rigid progression, since different materials can be most effective in different schemes, testing can result in changes to a design, and a final design is often reached by starting with a rough estimated design, then looping through several cycles of analysis and redesign. Often, several alternative designs will prove quite close in cost, strength, and serviceability. The structural engineer, owner, or end user would then make a selection based on other considerations.

1. Project requirements

Before starting design, the structural engineer must determine the criteria for acceptable performance. The loads or forces to be resisted must be provided. For specialized structures, this may be given directly, as when supporting a known piece of machinery, or a crane of known capacity. For conventional buildings, building codes adopted on a municipal, county, or state level provide minimum design requirements for live loads (occupants and furnishings, snow on roofs, and so on). The engineer will calculate dead loads (structure and known, permanent installations) during the design process.

<u>For the structural to be serviceable or useful, deflections must also be kept within limits, since it is possible for safe structures to be uncomfortably "bounce".</u>[①] Very tight deflection limits are set on supports for machinery, since beam sag can cause drive shafts to bend, bearings to burn out, parts to misalign, and overhead cranes to stall. Limitations of sag less than span/1000 (1/1000 of the beam length) are not uncommon. In conventional buildings, beams supporting ceilings often have sag limits of span/360 to avoid plaster cracking, or span/240 to avoid occupant concern (keep visual perception limited). Beam

stiffness also affects floor "bounciness", which can be annoying if not controlled. In addition, lateral deflection, sway, or drift of tall buildings is often held within approximately height/500 (1/500 of the building height) to minimize the likelihood of motion discomfort in occupants of upper floors on windy days.

Member size limitations often have a major effect on the structural design. For example, a certain type of bridge may be unacceptable because of insufficient under clearance for river traffic, or excessive height endangering aircraft. In building design, ceiling heights and floor-to-floor heights affect the choice of floor framing. Wall thicknesses and column sizes and spaces may also affect the serviceability of various framing schemes.

2. Materials selection

Technological advances have created many novel materials such as carbon fiber-and boron fiber-reinforced composites, which have excellent strength, stiffness, and strength-to-weight properties.② However, because of the high cost and difficult or unusual fabrication techniques required, they are used only in very limited and specialized applications. Glass-reinforced composites such as fiberglass are more common, but are limited to lightly loaded applications. The main materials used in structural design are more prosaic and include steel, aluminum, reinforced concrete, wood, and masonry.

3. Structural schemes

In an actual structure, various forces are experienced by structural members, including tension, compression, flexure (bending), shear, and torsion (twist). However, the structural scheme selected will influence which of these forces occurs most frequently, and this will influence the process of material selection.

Tension is the most efficient way to resist applied loads, since the entire member cross section is acting to full capacity and buckling is not a concern.③ Any tension scheme must also include anchorages for the tension members. In a suspension bridge, for example, the anchorages are usually massive dead weights at the ends of the main cables. To avoid undesirable changes in geometry under moving or varying loads, tension schemes also generally require stiffening beams or trusses.

Compression is the next most efficient method for carrying loads. The full member cross section is used, but must be designed to avoid buckling, either by making the member stocky or by adding supplementary bracing. Domed and arched buildings, arch bridges, and columns in building frames are common schemes. Arches create lateral outward thrusts which must be resisted. This can be done by designing appropriate foundations or, where the arch occurs above the roadway or floor line, by using tension members along the roadway to tie the arch ends together, keeping them from spreading. Compression members weaken drastically when loads are not applied along the member axis, so moving, variable, and unbalanced loads must be carefully considered.

Schemes based on flexure are less efficient than tension and compression, since the flexure or bending is resisted by one side of the member acting in tension while the other side acts in compression. Flexural schemes such as beams, girders, rigid frames, and moment (bending) connected frames have advantages in requiring no external anchorages or thrust restraints other than normal foundations, and inherent stiffness and resistance to moving, variable, and unbalanced loads.

Trusses are an interesting hybrid of the above schemes. They are designed to resist loads by spanning in the manner of a flexural member, but act to break up the load into a series of tension and compression forces which are resisted by individually designed tension and compression members. Truss schemes can be designed to require no special anchorage or thrust restraints and have excellent stiffness and resistance to moving and variable loads. Numerous member-to-member connections, supplementary compression braces, and a somewhat cluttered appearance are truss disadvantages.

Plates and shells include domes, arched vaults, sawtooth roofs, hyperbolic paraboloids, and saddle shapes. Such schemes attempt to direct all force along the plane of the surface, and act largely in shear. While potentially very efficient, such schemes have very strict limitations on geometry and are poor in resisting point, moving, and unbalanced loads perpendicular to the surface.

Stressed-skin and monocoque construction uses the skin between stiffening ribs, spars, or columns to resist shear or axial forces.④ Such design is common in airframes for planes and rockets, and in ship hulls. It has also been used to advantage in buildings. Such a design is practical only when the skin is a logical part of the design and is never to be altered or removed.

For bridges, short spans are commonly girders in flexure. As spans increase and girder depth becomes unwieldy, trusses are often used, as well as cablestayed schemes. Longer spans may use arches where foundation conditions, underclearance, or headroom requirements are favorable. The longest spans are handled exclusively by suspension schemes, since these minimize the crucial dead weight and can be erected wire by wire.

For buildings, short spans are handled by slabs in flexure. As spans increase, beams and girders in flexure are used. Longer spans require trusses, especially in industrial buildings with possible hung loads. Domes, arches, and cable-suspended and air-supported roofs can be used over convention halls and arenas to achieve clear areas.

4. Structural analysis

Analysis of structures is required to ensure stability (static equilibrium), find the member forces to be resisted, and determine deflections. It requires that member configuration, approximate member sizes, and material properties be known or assumed. Aspects of analysis include: equilibrium; stress, strain, and elastic modulus; linearity; plasticity; and curvature and plane sections. Various methods are used to complete the analysis.

5. Final design

Once a structure has been analyzed (by using geometry alone if the analysis is determinate, or geometry plus assumed member sizes and materials if indeterminate), final design can proceed. Deflections and allowable stresses or ultimate strength must be checked against criteria provided either by the owner or by the governing building codes. Safety at working loads must be calculated. Several methods are available, and the choice depends on the types of materials that will be used.

Pure tension members are checked by dividing load by cross section area. Local stresses at connections, such as bolt holes or welds, require special attention. Where axial tension is combined with bending moment, the sum of stresses is compared to allowable levels. Allowable stresses in compression members are dependent on the strength of material, elastic modulus, member slenderness, and length between bracing points. Stocky members are limited by material strength, while slender members are limited by elastic buckling.

Design of beams can be checked by comparing a maximum bending stress to an allowable stress, which is generally controlled by the strength of the material, but may be limited if the compression side of the beam is not well braced against buckling.

Design of beam-columns, or compression members with bending moment, must consider two items. First, when a member is bowed due to an applied moment, adding axial compression will cause the bow to increase. In effect, the axial load has magnified the original moment. Second, allowable stresses for columns and those for beams are often quite different. To reflect this, the following interaction equation is used. The member is considered satisfactory if:

$$\frac{\text{Actual axial}}{\text{Allowable axial}} + \frac{\text{magnified actual moment}}{\text{allowable moment}} \leqslant 1.0$$

Members that are loaded perpendicular to their long axis, such as beams and beam-columns, also must carry shear. Shear stresses will occur in a direction to oppose the applied load and also at right angles to it to tie the various elements of the beam together. They are compared to an allowable shear stress. These procedures can also be used to design trusses, which are assemblies of tension and compression members. Lastly, deflections are checked against the project criteria using final member sizes.

Once a satisfactory scheme has been analyzed and designed to be within project criteria, the information must be presented for fabrication and construction. This is commonly done through drawings, which indicate all basic dimensions, materials, member sizes, the anticipated loads used in design, and anticipated forces to be carried through connections.

Ⅰ. New Words

1. stationary *adj.* 静止的,不变的,固定的

2. hull *n.* 船壳，船体，船身

3. progression *n.* 级数；进展，连续

4. loop *n.* 环，圈，弯曲部分；循环

5. occupant *n.* 占有人，居住者

6. driveshaft *n.* 传动轴，从动辊，驱动杆

7. sag *vi.* 下垂，下降；*n.* 挠度

8. boron *n.* ［化］硼

9. fiberglass *n.* 玻璃纤维，玻璃丝

10. stocky *adj.* 矮胖的，短粗的，结实的

11. drastically *adj.* 激烈的，彻底的，急剧的

12. inherent *adj.* 固有的，内在的，与生俱来的

13. hybrid *n.* 混合，合成，混合物

14. dome *n.* 穹顶，圆屋顶，圆盖

15. saddle *n.* 鞍，鞍状物，鞍状构造

16. unwieldy *adj.* 笨重的，难使用的

17. headroom *n.* 净空，净空高度

18. underclearance *n.* 桥下净空，下部间隙

Ⅱ. Phrases and Expressions

1. structural scheme   结构方案
2. floor line   地面标高，地面标线
3. flexural member   受弯构件
4. cross section   横断面，横截面
5. sawtooth roof   锯齿形屋顶
6. fiber-reinforced concrete   纤维增强混凝土
7. hyperbolic paraboloid   双曲抛物面
8. saddle shapes   马鞍形
9. stiffening ribs   加劲肋
10. elastic modulus   弹性模量

Ⅲ. Notes

① "deflection" 在此句可译为 "挠度"。因为 "deflection" 作主语，所以谓语动词用被动语态，"be kept within limits" 表示在限值范围内。

专业英语文章的特点之一就是被动语态的广泛使用，原因是被动语态句比主动语态少主观色彩，更能客观地反映事实。但是在汉语表达中很少使用被动语态，因此，在翻译被动语态句型时，可根据具体的语言环境把被动语态译为主动语态形式。

② "Technological advances" 技术的进步，"advance" 译为 "进步"，同类表示词语还有 improvement，progress 等。

③ "the most efficient way" 是 "efficient" 的最高级表示形式，译为 "最有效的方

式","concern"译为"关心、担心",但"not a concern"可译为"不必担心"。

④ "skin"在此句译为"壳板","the skin between stiffening ribs, spars, or columns"译为"加劲肋、杆、柱之间的壳板"。

Ⅳ. Exercises

Fill in the blanks with the information given in the text.

1. Structural design involved at least five distinct phases of work: _____, _____, _____, _____, _____.

2. Member size limitations often have a major effect on the _____.

3. _____, _____, _____, and _____ are common schemes.

Ⅴ. Expanding

Know about the following terms related to the structural design.

1. limit state   极限状态
2. bearing capacity limit state   承载能力极限状态
3. serviceability limit state   正常使用极限状态
4. reliability   可靠性
5. reliable probability   可靠概率
6. failure probability   失效概率

Ⅵ. Reading Material

Steps in Structural Design Process

There are certain logical steps into which the design process falls. Although there are no hard-and-fast rules in this regard, the following factors are considered, and usually in the order indicated:

1. Functional requirements

It is essential first to establish the purpose for which the structure is to be used. This is especially true if its size, form, or use departs from the usual pattern that has seen service in the past. An example would be in the design of a large structure to house launching rockets where the requirement to operate doors under all circumstances places unusual deformation limits on the structure. Another aspect of their performance requirement is esthetics. Civil engineering works by their very nature are always in the public view and frequently tend towards the monumental; they always contribute to and sometimes dominate their surroundings. It is thus essential that the appearance of the structure be carefully considered; indeed, there are times when esthetics may constitute a major design criterion, as in the case of a bridge at a point of scenic interest. An essential part of the functional requirement is the consideration of how the structure relates to the larger system of which it is a part.

2. Structure and loading

Hand in hand with the functional requirement goes the development of the structural form to be used. The types of loads acting on the structure also are considered at this stage since the load and the structure are often mutually dependent. An example is the amplification of wind effects due to aerodynamic response of the structure. The fire loading and the method of protecting steel from the effects of fire would be examined in this phase. Along with the functional requirement and the development of structural form to support the loads goes the necessary consideration of the cost of the structure. It is an essential aspect of structural design.

3. Loading conditions

The selection of loading conditions follows the determination of the type and magnitude of the loads that act on the structure. It involves, in the main, a decision as to which loads act in combination with the dead load. Some codes specify the minimum loading conditions that must be considered, but it is part of the designer's basic responsibility to determine the extent and magnitude of the actual loading conditions. Occasionally, the structural form has a very significant influence on the loading conditions that must be considered. Again, where the form is unusual, more attention must be given to this feature.

4. Preliminary design

The initial selection of member sizes is called "preliminary design". As indicated earlier, it is a necessary step in allowable stress design, since the initial estimate of section size must be checked to see if the allowable stress is exceeded. In the plastic design of continuous beams and industrial building frames, this step frequently involves no more than a determination of the ratios of required member sizes.

5. Analysis

With the information obtained in the previous step, an analysis (theoretical, model, or full-scale experimental) is made to determine points of critical moment, force, or stress, from which are calculated the required design properties of each member.

6. Selection of section

This step consists of the final or revised selection of size and shape of members to be used in the design. The selection is based on the analysis of structural strength which leads to the required section properly—which could be the section area, the section modulus, or the plastic modulus.

7. Secondary design items

Prior to completing a design, a most important final step is to check on such factors as shear, local buckling, bracing, column buckling; to design connections; and, where required, to examine deflections, stiffness requirements, and any special material and design provisions to obviate brittle fracture under adverse conditions.

Some of the steps listed and described above may be bypassed, and frequently the step-by-step sequence will need to be modified. For example, in the interests of meeting an immediate objective which involves the national interest it may be necessary to give less attention to finding the most economical design and more attention to some other aspect. It is in a circumstance such as this that reliance must be placed upon judgment—a quality that is gained through experience, but which will have a better basis when founded on an understanding of structural behavior and systems. It is in part for this reason that such emphasis is laid in this book on the behavior of structures throughout the entire range of structural response: elastic, inelastic, and failure.

参 考 译 文

第12课 结 构 设 计

结构设计是选择材料和构件的类型、尺寸和形状，以安全耐久的方式来承担荷载。通常，结构设计是指对固定工程如建筑和桥梁的设计，或是虽可移动但却有一刚性外形的物体，如船体和飞机框架。设备与零配件被设计成互相联动通常是机械设计的范畴。

结构设计至少包括五个明确的工作阶段：项目要求、材料、结构方案、分析、设计。对于特殊的结构和材料，第六阶段——试验也应包括在内。这些阶段并非以严格不变的程序进行，因为不同的材料可以在不同的结构方案中发挥最大效能，而试验会导致设计变更，且最终的设计经常是由初步估算设计开始，再经过分析与重新设计的数次循环后才能完成。通常是，几个可供选择的设计在造价、强度和适用性等方面是非常接近的，结构工程师、业主或最终使用者再依据其他因素做出选择。

1. 项目要求

开始设计前，结构工程师必须确定可以接受的性能标准，应提供结构需抵抗的荷载或抗力。对于特定的结构，这些可直接给出，如结构需支承已知的机械设备或已知起重量的吊车。对于常规建筑，可采用地方、县和国家标准的建筑规范，对活荷载给出了设计所需的最小值（居住者和家具、屋面雪荷载等）。在设计过程中，工程师还要计算恒载（结构自重、已知的永久设备）。

为了满足结构的耐久性和适用性的要求，挠度必须保持在限值之内，因为对于安全的结构来说，也可能出现令人不安的震动。对机械设备支座变形的限制是十分严格的，因为梁下沉会导致驱动轴弯曲，烧坏，零部件移位和上面的吊车熄火。小于跨度1/1000的挠度限值并不罕见，在传统的建筑中，支撑楼板的梁为了避免抹灰开裂，其挠度限制为跨度的1/360，为避免使用者的关注，其挠度限值为跨度的1/240（避免视觉可见）。梁的刚度也会影响楼板的震动，如果不控制，这种震动会令人头疼。此外，高层建筑的侧向变形、摆动和位移通常要控制在大约为高度的1/500范围内，从而减小上部楼层的使用者在大风天时的不适位移。

在进行结构设计时，杆件尺寸限制是一个主要的影响因素。例如，某一类型的桥梁会由于河内运输的净空不足或高度过大而威胁到飞机等原因而不被采纳。在建筑设计中，顶棚高度和楼板到楼板的高度都将影响楼板框架的选择。墙厚和柱的尺寸及间距也可能影响各种框架方案的适用性。

2. 材料选择

科技的进步创造了许多新材料，比如碳纤维和硼纤维增强复合材料，它们都具有极好的强度、刚度和强度重量比等特性。然而，由于价格较高，需要较难或不寻常的加工技术，它们仅在非常有限和特殊的领域使用。玻璃增强复合材料，比如玻璃纤维更为通用，但仅限于荷载较小的应用范围。用于结构设计中的主要材料多数都是普通的材料，包括钢材、铝、钢筋混凝土、木材及砌体。

3. 结构方案

在实际结构中，结构杆件要受到各种力的作用，包括拉力、压力、弯矩、剪力和扭矩。然而，所选择的结构方案将影响这些力的产生类型，也会影响材料的选择过程。

受拉是抵抗外荷载最有效的方式，因为杆件全截面可以发挥最大承载力，而不用担心压曲。任何一种受拉方案必须包括受拉杆件的锚固。比如在悬索桥中，锚点通常在主要悬索端部且大量恒载集中点。为了避免在移动和可变荷载下产生几何变形，受拉设计通常要求加强梁或桁架。

受压是承担荷载的另一种有效方法。杆件全截面被充分利用，但设计时必须避免压曲，可以将杆件设计成短粗构件或增加辅助支撑。圆形和拱形建筑、拱桥和结构框架中柱是普通的方案。拱产生需要抵抗的外部水平推力。这种推力可以设计适当的基础来平衡，或当拱位于路面以上或楼板面以上时，用拉杆沿着路面将拱端部连在一起，以阻止拱端部分开。当荷载不作用在杆件轴线上时，抗压能力会迅速降低，所以必须仔细考虑移动、多变和不平衡荷载。

抗弯要比受拉和受压效果差一些，因为抵抗弯矩时，杆件截面一侧受拉，同时另一侧受压。受弯杆件如梁、大梁、刚架和连接框架的节点有其自己的优点，即不需要外部锚固或除了正常基础以外的反向推力，有其固有刚度和抵抗移动、可变和不平衡荷载的能力。

桁架是上述几种方案的有利组合。它们被设计成以避免受弯构件的方式抵抗荷载，但把荷载分成一系列由单个拉杆和压杆抵抗的拉力和压力。设计桁架时不需要锚固或者反力约束，并具有极好的刚度和抵抗可变荷载的能力。桁架的缺点是有许多杆件间的连接和附加受压支撑，而且外观有些凌乱。

板和壳包括圆形、拱形、锯齿形屋顶、双曲抛物面和马鞍形状。这种方案试图直接在平面内受力，并承担很大剪力，而且非常有效，但这种方案在几何尺寸方面限制非常严格，并且不利于抵抗垂直于平面的集中、移动和不稳定荷载。

薄壳结构和硬壳结构利用加劲肋、梁和柱之间的壳板抵抗剪力和轴力。这种设计在飞机、火箭的机体和轮船船身中常见。在建筑设计中也可以发挥其优势。当薄壳在设计中是合理的一个组成部分，并永不改变和移动时，这种方案才非常适用。

在桥梁中，短跨的弯曲大梁比较常见。当跨度增大时，梁高就会随之增大而变得难于使用，这时可以采用桁架或拉索方案。当基础情况、桥下净空和上部空间允许时，大跨可以采用拱方案。当跨度最大时，可以专门采用悬索方案，因为这些悬索可以把至关重要的恒载降低到最低限值，并且可以用一根接一根的钢绞线组建而成。

在建筑结构中，短跨时可以采用板抗弯；当跨度增加时，可以使用梁和大梁抗弯；当跨度再大时，尤其是在可能有悬挂荷载的工业建筑中，可以采用桁架结构。对于要获得较大净空面积的大会堂和舞台，可以采用圆形、拱形、悬索和气撑式屋顶。

4. 结构分析

结构分析要求确保结构稳定（即静力平衡），找到构件的抗力，以及确定挠度，这要求已知或假定构件的形状、构件的近似尺寸和材料性能。分析方面还包括：平衡；应力、应变和弹性模量；线性（分析）；塑性（分析）；曲率与平截面。可运用各种方法完成分析。

5. 最终设计

一旦结构分析完（若结构是静定的可单独运用几何学的方法；若结构是静不定的，则要运用几何学加假定构件尺寸和材料的方法），就可进行最终设计。挠度和允许应力或极限强度必须经得住业主或国家建筑法规提供的标准的检验，必须计算出工作荷载下的安全性，有几种方法可以使用，选择哪种方法取决于所用材料的类型。

单纯的受拉构件可用荷载除以横截面面积检验。连接处的局部应力，如螺栓孔或焊缝，要特别注意。在轴心受拉与弯矩结合处，应力总和不能超过允许值。抗压构件的允许应力取决于材料的强度、弹性模量、长细比和支点间的长度。短粗构件受材料强度的限制，而细长构件却受弯曲弹性的限制。

梁的设计可用最大弯曲应力除以允许应力来检验，这通常受材料强度的限制，但是如果梁的受压边不能很好地固定住以抵抗弯曲，也可能受限制。

梁-柱的设计，或有弯矩的受压构件的设计，必须考虑以下两条：第一，当由于施加弯矩使构件弯曲时，增加轴心受压将增大弯曲，实际上，轴心荷载增大了初始弯矩。第二，柱中的允许应力和梁的允许应力是完全不同的，为反映这一不同，可应用下面的公式，如果：

$$\frac{实际轴向力}{允许轴向力}+\frac{放大的实际力矩}{允许力矩}\leqslant 1.0$$

可认为构件是合适的。

垂直于长轴加载的构件，如梁和梁-柱，也必须能承担剪力。在承担加载方向将产生剪应力，在垂直于加载方向也将产生剪应力，将梁的各个部分联系在一起。它们相当于允许剪应力。这些步骤也可用来设计受拉和受压构件组合而成的桁架。最后，对照设计标准用构件的最终尺寸来检算挠度。

当一个合适的方案在设计标准范围内分析和设计完成之后，就应为制作和施工提供资料。这通常依据图纸进行，图纸中指出了所有的基本量纲、材料、构件尺寸、设计中使用的预期荷载及通过连接要承担的预期力。

Lesson 13
Masonry Structure

Masonry is one of man's oldest building materials and probably most easily understood. Such misconceptions have led over the years to a serious misuse of the material through inadequate or even nonexistent design procedures and poor construction practices. However, perhaps because of the considerable amount of information and data available today, both as to its properties and structural performance, sound design techniques and vastly improved construction practices have evolved within recent years, all of which make for optimum use of the material's capabilities.① This is in no small way due to the effort continually being exerted toward this evolution by such diverse agencies as the International Conference of Building Officials (ICBO) and the Masonry Institute of America (MIA).

Masonry is a totally different and distinct type of construction material, not one that is "sort of like reinforced concrete". It is not, and should not, be treated as such. Furthermore, the wind, seismic, and structural performance research carried on during the recent past has resulted in building codes of increasing complexity.② This, in turn, has led to more sophisticated and comprehensive methods of design.

Masonry is primarily a hand-placed material whose performance is highly influenced by factors of placement. Hence, knowledge of the basic ingredients (i.e. mortar, grout, masonry unit, and reinforcement) is essential if a practical and efficient design conception is to be achieved. In addition, if the design is to be brought to a successful fruition, as its designer conceived it, proper inspection procedures must be followed to ensure that its delivery will be more certain. Furthermore, before anyone can hope to turn out an adequate design of any sort, he or she must possess a rudimentary knowledge of the properties and performance of the materials being employed.

Next in the process comes the need for a description of the various load sources and intensities, a presentation of the fundamental precepts, and the development of the very basic design and analysis expressions as they evolve from the basic structural mechanics without reference to code limitations or empirical rules. The many code requirements must then be incorporated into these basic expressions and relations to produce an integrated design procedure,③ one that will result in very practical solutions to the engineering problems normally encountered by structural engineers in everyday practice.

The total design of a masonry building begins with a consideration of the preliminary and nonstructural aspects of masonry bearing on the case study, such as its fire-resistive or environmental features. Following this examination comes the determination of the live, dead, seismic, and wind loads—their magnitudes and stress paths from point of application to the ground. Finally, the member sizes and reinforcing requirements are selected, adequate connections are devised, and the system is detailed such that it can be readily constructed. The latter is an extremely important consideration, but one too often slighted or ignored. In the past, this total concept has not been given the emphasis it deserves. Many textbooks seem to ignore the aspect of stability of the total framing system as it resists lateral loads, focusing instead on the behavior of the individual beams, columns, walls, and other elements comprising the system. Certainly, modern buildings almost everywhere are subjected to significant lateral loads of one type or another to varying degrees of magnitude. The placement of the entire country into seismic zones of various degrees of probability and intensity has only served to accentuate this critical factor. To ignore it is folly, as some have found to their chagrin. It really is not overly difficult to design a building to withstand gravity loads. But developing a lateral-force-resisting system (frame, shear walls, or combination thereof) requires skill and imagination—a process that taxes the ingenuity of structural engineers to come up with solutions that are in all ways safe, practical, and yet economical.

Brick is actually the oldest manufactured building material remaining in use today. In the pre-modern era, the development of brick masonry reached its fruition in the United States and Europe. The successful use of this ancient material is certainly demonstrated in many early American brick structures, such as the Monadnock Building in Chicago. But its very massiveness discouraged further use of unreinforced masonry bearing walls for high-rise buildings. This condition remained unchanged for nearly 50 years, awaiting the advent of modern reinforced masonry. The Monadnock represented the watershed, in America at least, of the use of plain masonry bearing walls.

The classifications of masonry construction and the types of masonry walls appear in the UBC. The distinction between these various categories must be thoroughly understood by anyone who intends to design masonry under UBC jurisdiction. For this reason, they are thoroughly delineated in the following sections.

Masonry construction is classified as follows: (1) "reinforced masonry", which must be engineered on the basis of sound theoretical principles combined with a set of empirical rules and limitations set forth by the Building Code, plus sound engineering judgment stemming from long experience; (2) "partially reinforced masonry", which was introduced into the Uniform Building Code primarily for those areas in which all the requirements of reinforced masonry were not needed, since the seismicity of the local did not so dictate; (3) "unreinforced engineered masonry", which was developed in the East as an attempt to improve on past practices, many of which were unsound; and (4) "traditional masonry", which encompasses the use of masonry as it evolved over the years from certain arbitrary limitations and past practices without any real consideration for theoretical design characteristics; although it did provide for a generally conservative and safe type of construction for the majority of conditions.

Lesson 13　Masonry Structure

Ⅰ. New Words

1. misconception *n.* 误解，错觉
2. optimum *adj.* 最佳的，最优的，最适宜的
3. exert *vt.* 运用，发挥，施加，尽力
4. seismic *adj.* 地震的
5. sophisticated *adj.* 复杂的，精密的，尖端的
6. comprehensive *adj.* 广泛的，综合的
7. conceive *vt.* 构想出，设想
8. rudimentary *adj.* 基本的，初步的
9. precept *n.* 规则，教训
10. magnitude *n.* 重要性，大小，量级
11. folly *n.* 愚蠢，荒唐
12. chagrin *n.* 失望，懊恼
13. ingenuity *n.* 心灵手巧，独创性
14. watershed *n.* 转折点，分水岭，分水线
15. delineate *vt.* 描绘，描写
16. empirical *adj.* 经验上的，经验主义的
17. arbitrary *adj.* 任意的，武断的

Ⅱ. Phrases and Expressions

1. structural performance　结构性能
2. masonry unit　砌块
3. lateral load　横向荷载
4. lateral-force-resisting system　抗侧力体系
5. live load　活载荷
6. dead load　恒载荷
7. unreinforced masonry　无筋砌体
8. reinforced masonry　配筋砌体

Ⅲ. Notes

① "all of which"引导了一个非限定性定语从句，起到解释和说明的作用。
② "furthermore"是表示递进关系的连接副词，译为"除此以外"。此外，同义的表示递进关系的副词或短语还有 moreover, what's more, in addition 等。
③ 词组"be incorporated into"的意思是"合并到，纳入"，也可以写成"be incorporated to"的形式。

Ⅳ. Exercises

Translate the following sentences into Chinese.

1. Next in the process comes the need for a description of the various load sources and intensities, a presentation of the fundamental precepts, and the development of the very basic design and analysis expressions as they evolve from the basic structural mechanics without reference to code limitations or empirical rules.

2. The total design of a masonry building begins with a consideration of the preliminary and nonstructural aspects of masonry bearing on the case study, such as its fire-resistive or environmental features.

3. The distinction between these various categories must be thoroughly understood by anyone who intends to design masonry under UBC jurisdiction.

Ⅴ. Expanding

Know about the following terms related to the masonry structure.
1. brick masonry   砖砌体
2. stone masonry   石砌体
3. rubble   毛石
4. solid brick   实心砖
5. hollow; cored brick   空心砖
6. porous brick   多孔砖
7. coal gangue hollow brick   煤矸石空心砖

Ⅵ. Reading Material

The Characteristics of the Masonry Structure

This article refers to the building structure component. Masonry is the building of structures from individual units laid in and bound together by mortar; the term masonry can also refer to the units themselves. The common materials of masonry construction are brick, stone such as marble, granite, travertine, limestone; concrete block, glass block, and tile. Masonry is generally a highly durable form of construction. However, the materials used, the quality of the mortar and workmanship, and the pattern in which the units are assembled can strongly affect the durability of the overall masonry construction.

Masonry units, such as brick, tile, stone, glass brick or concrete block generally conform to the requirements specified in the 2006 International Building Code (IBC) Section 2103.

1. Applications

Masonry is commonly used for the walls of buildings, retaining walls and monuments. Brick and concrete block are the most common types of masonry in use in industrialized nations and may be either weight-bearing or a veneer. Concrete blocks, especially those with hollow cores, offer various possibilities in masonry construction. They generally provide great compressive strength, and arc best suited to structures with light transverse loading when the cores remain unfilled. Filling some or all of the cores with concrete or concrete with steel reinforcement (typically rebar) offers much greater tensile and lateral strength to structures.

2. Advantages

(1) The use of materials such as brick and stone can increase the thermal mass of a building, giving increased comfort in the heat of summer and the cold of winter, and can be ideal for passive solar applications.

(2) Brick typically will not require painting and so can provide a structure with reduced life-cycle costs, although sealing appropriately will reduce potential spalling due to frost damage. Non-decorative concrete block generally is painted or stuccoed if exposed.

(3) The appearance, especially when well crafted, can impart an impression of solidity and permanence.

(4) Masonry is very heat resistant and thus provides good fire protection.

(5) Masonry walls are more resistant to projectiles, such as debris from hurricanes or tornadoes than walls of wood or other softer, less dense materials.

3. Disadvantages

(1) Extreme weather causes degradation of masonry wall surfaces due to frost damage. This type of damage is common with certain types of brick, though rare with concrete block. If non-concrete (clay-based) brick is to be used, care should be taken to select bricks suitable for the climate in question.

(2) Masonry tends to be heavy and must be built upon a strong foundation (usually reinforced concrete) to avoid settling and cracking. If expansive soils (such as adobe clay) are present, this foundation needs to be quite elaborate and the services of a qualified structural engineer may be required, particularly in earthquake prone regions.

4. Structural limitations

Masonry boasts an impressive compressive strength (vertical loads) but is much lower in tensile strength (twisting or stretching) unless reinforced. The tensile strength of masonry walls can be strengthened by thickening the wall, or by building masonry piers (vertical columns or ribs) at intervals. Where practical, steel reinforcements can be added.

参 考 译 文

第13课 砌 体 结 构

砌体是人类最古老的建筑材料之一，并且可能最容易被理解掌握。多年来，这种误解导致通过不适当的或甚至不存在的设计程序和糟糕的施工实践严重滥用了该材料。然而，大概是因为当今我们能获得大量的有关砌体性质和结构性能的信息及数据，近年来，合理的设计技术和大量改进的施工实例不断涌现，这些都是为了达到材料性能的最优使用。这是由于诸如国际建筑官员大会（ICBO）和美国砌体研究所（MIA）等不同机构对这种转变的不断努力的结果。

砌体是一种完全不同的和独特的建筑材料，不是"有几分像钢筋混凝土"的建筑材料。它不是，也不应该被认为是这样。此外，近年来对风荷载、地震力以及结构性能的研究已经导致建筑规范越来越复杂。这反过来也导致设计的方法更复杂、更全面。

砌体主要是手工方式砌筑的材料，其性能受到砌筑因素的高度影响。因此，如果要实现实用和有效的设计概念，则必须了解基本组成材料（即砂浆、水泥浆、砌体块材和钢筋）。此外，如果设计要得以成功实现，如设计者设想的那样，必须遵循适当的检查程序，以确保其交付更加有把握。此外，在任何人都希望设计出适当的设计之前，他或她必须对所使用的材料的性质和性能具有初步的了解。

在接下来的过程中需要描述各种荷载的来源和强度，介绍基本规律，以及最基本的设计和分析表达式的发展，这些表达式从基本结构力学逐步演变而成但未参考规范限制或经验规则。然后，许多规范要求必须被纳入到这些基本的表达式和关系式当中，以形成一个完整的设计过程，这对于结构工程师在日常实践中通常遇到的工程问题将产生非常实用的解决方案。

砌体结构建筑的总体设计起始于对方案研究的初步考虑和非砌体结构承重方面的考虑，例如其耐火性能或环境特点。接下来该检查确定活荷载、恒荷载、地震力和风荷载，确定它们的大小和从加载位置到地面的荷载传递路径。最后，选择构件尺寸和抗力要求，设计可靠的连接，并且该连接体系要被详细地说明，使其可以容易地构建。后者是一个非常重要的考虑因素，但往往被轻视或忽视。在过去，这个总的概念设计没有得到应有的重视。许多教科书似乎忽视了整个结构体系在抵抗横向载荷时的稳定性方面，而是集中于单独的梁、柱、墙和组成系统的其他构件的行为。当然，几乎所有地方的现代建筑物都受到一种或另外一种不同程度横向荷载的明显作用。将整个国家置于不同程度的概率和烈度的地震区中只是突出了这一关键因素。忽略它是愚蠢的，因为有些人已经因此而懊悔。设计一个承受重力荷载的建筑真的不是太难。但是，开发一种抗侧力系统（框架、剪力墙或框架剪力墙）是需要技能和想象力的过程，这是对结构工程师的聪明才智的考验，以提出在所有方面安全、实用、经济的解决方案。

砖实际上是目前仍在使用的最古老的人造建筑材料。在前现代时期，砖砌体的发展在美国和欧洲取得了成果。这种古老材料的成功使用无疑在许多早期的美国砖结构中得到证明，如芝加哥的莫纳德诺克大厦。但它的庞大自重阻碍了无筋砌体承重墙在高层建筑的进一步使用。这种情况在近 50 年内保持不变，一直到现代配筋砌体的出现。至少在美国，莫纳德诺克大厦代表了无筋砌体承重墙使用的分水岭。

在 UBC 中出现了砌体结构和砌体墙类型的分类。任何想在 UBC 管辖下设计砌体的人都必须深入了解这些不同类别之间的区别。因此，在下面的章节中对它们进行了详细的描述。

砌体结构分为以下几类：（1）"配筋砌体"，它必须在完善的理论原理并结合建筑规范规定的一套经验规则和限制的基础上设计，加上来自长期经验的合理工程判断；（2）"部分配筋砌体"，它被引入到统一建筑规范中，主要用于不需要配筋砌体的那些地区，因为当地的地震活动没有这样的规定；（3）"无筋砌体工程"，这是在东方发展起来的作为改善过去做法的一种尝试，其中有许多是不健全的；（4）"传统砌体"，其中包括砌体的使用也是不健全的，因为它从多年来某些主观限制和过去的做法演变而来，没有真正考虑任何理论设计的特点，尽管它为大多数情况提供了一般保守和安全的结构类型。

Lesson 14
Reinforced Concrete Structure

Reinforced concrete is by reason of its strength, durability, availability and adaptability an economical material eminently suited for many types of permanent structures. ① Concrete is relatively strong in compression and can be made sufficiently massive to provide lateral stability. Steel, on the other hand, is strong in tension but often lacks adequate lateral stability in compression due to its slender proportions. A reinforced concrete member in which the concrete resists compression while the steel resists tension is therefore an ideal structural partnership.

Concrete is strong in compression but weak in tension. As a result, cracks develop whenever loads, or restrained shrinkage or temperature changes, give rise to tensile stresses in excess of the tensile strength of the concrete. In the plain concrete beam, the moments due to applied loads are resisted by an internal tension-compression couple involving tension in the concrete. ② Such a beam fails very suddenly and completely when the first crack forms. In a reinforced concrete beam, steel bars are embedded in the concrete in such a way that the tension forces needed for moment equilibrium after the concrete cracks can be developed in the bars. The construction of a reinforced concrete member involves building a form or mold in the shape of the member being built. The form must be strong enough to support the weight and hydrostatic pressure of the wet concrete, and any forces applied to it by workers, concrete buggies, wind, and so on. The reinforcement is placed in this form and held in place during the concreting operation. After the concrete has hardened, the forms are removed.

Portland cement concrete is, sufficiently impervious to moisture, and retains sufficient alkalinity, to inhibit corrosion of the reinforcement even when small cracks are present. Although the idea of reinforced concrete was suggested in 1830, the earliest successful use of tensile reinforcement in Portland cement concrete is believed to be due to the French contractor J. L. Lambot. He used a mesh of iron bars to build a reinforced concrete rowing boat for the Paris International Exhibition of 1855. An Englishman, William B. Wilkinson of Newcastle upon Tyne, took out the first patent for reinforced concrete in 1854, which proposed embedding either flat iron rods or second-hand wire ropes in the tension zones of concrete beams. This patent revealed that Wilkinson also understood the need to develop good bond between the concrete and the reinforcement, and the fact that concrete floors possessed good fire resistance. Wilkinson, together with Joseph Monier, a French engineer who later took out several patents, are generally regarded as the inventors of reinforced concrete.

The widespread use of concrete in engineering construction stems from its cheapness compared with other structural materials currently available. Its lack of tensile strength is overcome by including reinforcement, usually in the form of steel bars, to produce a composite material known as reinforced concrete. <u>Although the steel reinforcement does not prevent cracking of the concrete in regions of tension, it does prevent the cracks from widening, and it provides an effective means for resisting the internal tensile forces.</u>③ The quantity of reinforcement needed is usually quite small, relative to the volume of concrete, so that the total cost of reinforced concrete construction remains commercially very competitive.

The properties of reinforced concrete depend upon those of concrete and steel but in a number of cases they do not coincide. Both of the bars hinders free contraction of the concrete and leads to incipient stresses in both components: tension in the concrete and compression in the bars. Such contraction stresses are minimized by shortening uninterrupted structural lengths through the use of joints. Concrete creep, which is caused by constant loads, redistributes stresses between the concrete and embedded bars and offers the possibility of making full use of strength of the two components in axially compressed members.

The primary purpose of the steel reinforcement is to carry internal tensile forces. Reinforcement is therefore placed in beams near the tensile face, i.e. near the lower face in the in-span regions of positive moment and near the upper face in regions near internal supports, where negative moments act. In reinforced concrete design, it is important to provide reinforcement in all regions of potential cracking. <u>Thus a rectangular arrangement of vertical and horizontal steel bars is introduced into regions of a beam where inclined cracks can form as a result of combined shearing action and bending moment.</u>④ The longitudinal steel, or main reinforcement, and the transverse bars, called stirrups, may be preassembled into a reinforcing cage for ease of construction.

In a floor slab, bars are usually laid in the two main span directions at right angles, to resist the tensile forces produced by the bending actions in each direction. For ease of construction, welded mesh is frequently used as slab reinforcement.

Although steel reinforcement is used primarily to carry the internal tensile forces produced by external loading, it also has other uses. Steel is much stronger than concrete in compression, and it is sometimes used to boost the resistance of zones of compression, when the overall dimensions of the member are restricted. Longitudinal steel is placed in all compression members. In such members, bending is usually present in addition to axial compression and the longitudinal steel reinforcement at each face of the member may act either in tension or in compression. Transverse ties are used to maintain the longitudinal steel in position during casting of the concrete and later to prevent its outward buckling when it is subjected to compressive stress. Again, the reinforcing steel for the column may be preassembled into a cage.

Cracking in concrete may be caused not only by external loading, but also by temperature gradients and differential or restrained shrinkage. Secondary reinforcement is therefore provided to control such cracking, which may be unsightly and even dangerous.

Ⅰ. New Words

1. availability *n.* 有效性，实用性，可用性
2. eminently *adv.* 突出地，显著地
3. impervious *adj.* 不受影响的，不透水的
4. inhibit *vt.* 阻止，妨碍，抑制
5. reinforcement *n.* 加强，加固
6. crack *vi.* 破裂，爆裂；*n.* 裂缝
7. coincide *vi.* 同时发生，相符，一致
8. hinder *vt.* 妨碍，阻碍，阻止
9. incipient *adj.* 早期的，初期的
10. rectangle *n.* 矩形，长方形
11. inclined *adj.* 倾斜的，趋于……的
12. transverse *adj.* 横断的，横的
13. stirrup *n.* 箍筋，马镫
14. cage *n.* 笼子
15. boost *vt.* 提高，促进
16. shrinkage *n.* 收缩，减少
17. unsightly *adj.* 难看的，不雅的

Ⅱ. Phrases and Expressions

1. give rise to   引起
2. in excess of   超过，多于
3. be embedded in   被嵌入，被埋入
4. Portland cement   波特兰水泥
5. composite material   复合材料
6. redistribute stresses   应力重新分配
7. reinforcing cage   钢筋笼
8. stem from   起源于
9. temperature gradient   温度梯度
10. be subjected to   使经受，使遭受

Ⅲ. Notes

① "by reason of its strength, durability, availability and adaptability" 是一个介词短语，翻译时可先抓主干而省略其他内容，然后再补充上去。"eminently suited for many types of permanent structures" 是过去分词短语作后置定语，修饰 "material"。

②"the plain concrete beam"译为"素混凝土梁",其中"plain"有"普通的,素的"等含义。

③"it does prevent the cracks from widening"中的"does"在此句中起强调作用,可译成"确实能限制裂缝的发展"。

④"be introduced into"译为"被引入,插入";"as a result of"译为"由于……的结果,作为……的结果"。

Ⅳ. Exercises

Translate the following sentences into Chinese.

1. Reinforced steel is also used in the compression zones of beams to help the concrete resist compression stresses where, for design reasons, this becomes necessary, and for the same reason in columns to help carry larger loads.

2. Concrete and steel work well when bonded together because the coefficients of thermal expansion are practically the same, being from 5.5×10^{-6}, to 7.5×10^{-6} for concrete and 6.5×10^{-6} for steel; thus temperature variations will not produce disruptive stresses between them.

3. The specified characteristic strength of the steel is the value of the yield stress below which shall fall not than 5% of the test results of the material supplied as complying with the requirements of this standard.

Ⅴ. Expanding

Know about the following terms related to the concrete structure.

1. frame structure   框架结构
2. shear wall structure   剪力墙结构
3. tube structure   筒体结构
4. plain concrete   素混凝土
5. concrete filled steel tube   钢管混凝土
6. steel reinforced concrete   型钢混凝土
7. composite structure   组合结构

Ⅵ. Reading Material

Bond between Concrete and Steel

When we say two different materials have concerted action in a structure, we mean that they have the same strain at their points of contact. Thus, two beams with one beam simply superimposed on the other do not act concertedly. But if shear connectors are inserted to lock them together so that they have the same strain along their faces of contact, then there will be concerted action between these two beams. For reinforced concrete, once the concrete has hardened, the embedded steel is firmly bonded in the concrete and the two materials will have the same deformation under load. So it is important to assure that this bond between concrete and steel is not destroyed.

Because of the excellent bond between concrete and steel, stress differences in the steel reinforcement can be balanced by the concrete through bond.

Fig. 14-1 shows a segment dx of reinforcing bar of diameter d embedded in concrete. If there were a stress difference of $d\sigma$ at both ends of the rod, then it would need an average bond stress of τ along the surface of this rod segment to keep the bar segment in equilibrium. This required bond stress is

$$\tau = \frac{d}{4}\frac{d\sigma}{dx}$$

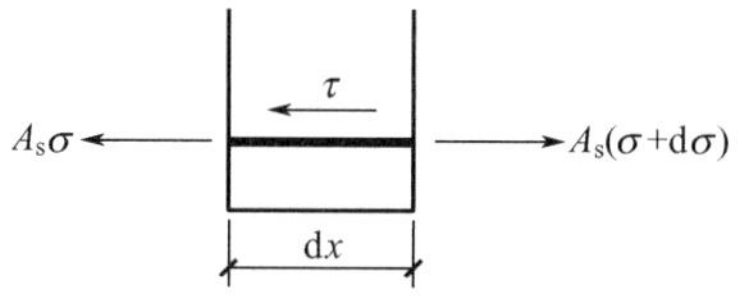

Fig. 14-1 Bond between concrete and steel rod

Unless the bond strength between concrete and steel is adequate to provide this required bond stress, the rod will slip in the concrete and the two materials will no longer act concertedly. Here it is important to note that for the same rate of change in stress, a thicker rod requires higher bond strength. So generally it is sensible to use a larger number of smaller bars rather than to use fewer number of larger bars to provide the same amount of reinforcement.

To increase the bond strength between concrete and reinforcement, deformed bars are developed with lugs or indentations rolled on them. For plain bars, bond strength is derived only from adhesion and inter-surface friction between concrete and steel. For deformed bars, bond strength is derived mainly from mechanical bearings of concrete on the lugs or indentations and thus has much higher bond strength. As it has been noted, in China, only grade I bars are plain bars, all other grade bars are deformed bars. It is also found that a thin coat of firm adhering rust on the rod is beneficial for bond requirement and the reason is obvious.

参 考 译 文

第 14 课　钢筋混凝土结构

钢筋混凝土因其强度大、耐久性好、实用性与适应性强而成为一种非常适合于许多永久性建筑物的经济的建筑材料。混凝土抗压相对较强，并且可以制成足够大以提供横向稳定性。另外，钢筋的抗拉很强，但由于其细长的比例，常常在受压时缺乏足够的横向稳定性。因此，钢筋混凝土构件中混凝土抵抗压力而钢筋抵抗拉力是一种理想的结构合作关系。

混凝土抗压强，但抗拉弱。因此，每当负载或有约束的收缩或温度变化引起的拉应力超过混凝土抗拉强度时，裂纹就会发展。在素混凝土梁中，因施加的载荷引起的力矩由内部的拉力与压力形成的力偶来抵抗，这个力偶中包含了混凝土中的拉力。当第一裂纹形成时，这种梁会非常突然而完全地破坏。在钢筋混凝土梁中，钢筋被埋入混凝土中，以这样的方式在钢筋中可以形成混凝土开裂之后弯矩平衡所需的拉力。钢筋混凝土构件的施工包括按照在建构件的形状建造模板或模具。该模板必须足够坚固以支撑湿混凝土的重量和流体静压，以及由工人、混凝土轨道车、风力等施加于其上的任何力。在混凝土浇筑操作期间钢筋放置在模板中并保持位置固定。混凝土硬化后，拆除模板。

波特兰水泥混凝土具有足够的防渗防潮性，并且保持足够的碱度，即使在存在小裂纹时也能抑制钢筋的腐蚀。虽然钢筋混凝土的想法是在1830年提出的，但最早在波特兰水泥混凝土中成功使用受拉钢筋的被认为是法国承包商 J.L. 兰波特。他用一个铁筋网为1855年的巴黎国际展览会建造了一个钢筋混凝土划艇。一个英国人，泰恩河畔纽卡斯尔的威廉·B. 威尔金森在1854年取得钢筋混凝土的第一个专利，提出在混凝土梁的受拉区埋入扁铁条或二手钢缆。该专利表明，威尔金森还理解了在混凝土和钢筋之间发展良好黏结的需要，以及混凝土楼板具有良好的耐火性的事实。威尔金森和法国工程师约瑟夫·莫尼尔，后来取得了几项专利，通常被视为钢筋混凝土的发明者。

混凝土在工程建设中的广泛应用，源于其与目前可用的其他结构材料相比较廉价。通过加入加强件，通常为钢筋的形式，来形成称为钢筋混凝土的复合材料，克服了其缺乏抗拉强度的问题。虽然钢筋不能防止混凝土受拉区裂缝的产生，但确实能限制裂缝的开展，因此提供了抵抗内部拉力的有效手段。所需的钢筋量相对于混凝土体积来说通常很少，使得钢筋混凝土结构的总成本在商业上保持着非常强的竞争力。

钢筋混凝土的性能取决于混凝土和钢的性质，但是在许多情况下它们是不一致的。钢筋两端阻碍混凝土的自由收缩并导致两部分材料内产生初始应力，即混凝土中的拉应力和钢筋中的压应力。通过使用施工缝缩短结构的连续长度，会使这种收缩应力最小化。由不变荷载引起的混凝土徐变在混凝土和内部钢筋之间产生应力重分布，并且提供了充分利用轴心受压构件中两种材料强度的可能性。

钢筋的主要作用是承担内部拉力。因此，钢筋被放置在梁的受拉区，即在正弯矩作用下的跨内区域中靠近下边缘，在负弯矩起作用的中间支座附近区域靠近上边缘。在钢筋混凝土设计中，重要的是在所有潜在开裂区域布置钢筋。因而，在由于剪力和弯矩的共同作用而使梁产生斜裂缝的区域放置呈矩形的竖向钢筋和水平钢筋。纵向钢筋、主筋和称为箍筋的横向钢筋，可以预先绑扎成钢筋笼以便于施工。

在楼板中，钢筋通常放置在沿直角的两个主跨方向上，以抵抗在每个方向上由弯曲作用产生的拉力。为了便于施工，板中的钢筋往往焊成网状。

虽然钢筋主要用于承受由外部荷载产生的内部拉力，但它也有其他用途。在受压时，钢筋比混凝土强得多，且当构件的总尺寸受到限制时，它有时用于提高受压区的抗力。纵向钢筋放置在所有受压构件中。在这样的构件中，除了轴向压缩之外通常存在弯曲，并且位于构件每一侧的纵向钢筋可以发挥作用，或者受拉或者受压。横向钢筋用于在混凝土浇筑期间将纵向钢筋保持在固定位置，随后在其承受压应力时防止其向外屈曲。此外，用于柱的钢筋可以预装成钢筋笼。

混凝土开裂不仅有可能是外部荷载引起的，而且有可能是温度梯度和差异或约束收缩引起。因此，利用间接钢筋来控制开裂，这可能是不美观的，甚至是危险的。

Lesson 15
Steel Members

1. Tension members

Tension members are found in bridge and roof trusses, towers, bracing systems, and in situations where they are used as tie rods. The selection of a section to be used as a tension member is one of the simplest problems encountered in design. As there is no danger of buckling, the designer needs only to compute the factored force to be carried by the member and divide that force by a design stress to determine the effective cross-sectional area required. Then it is necessary to select a steel section that provides the required area. Though these introductory calculations for tension members are quite simple, they do serve the important tasks of getting students started with design ideas and getting their "feet wet"[①] regarding the massive LRFD[②] Manual.

One of the simplest forms of tension members is the circular rod, but there is some difficulty in connecting it to many structures. The rod has been used frequently in the past, but has only occasional uses today in bracing systems, light trusses, and in timber construction. One important reason rods are not popular with designers is that they have been used improperly so often in the past that they have a bad name; however, if designed and installed correctly, they are satisfactory for many situations.

The average size rod has very little stiffness and may quite easily sag under its own weight, injuring the appearance of the structure. The threaded rods formerly used in bridges often worked loose and rattled. Another disadvantage of rods is the difficulty of fabricating them with the exact lengths required and the consequent difficulties of installation.

When rods are used in wind bracing it is a good practice to produce initial tension in them, as this will tighten up the structure and reduce rattling and swaying. To obtain initial tension the members may be detailed shorter than their required lengths, a method that gives the steel fabricator very little trouble. A common rule of thumb used is to detail the rods about $\frac{1}{16}$ inches short for each 20 feet of length { approximate stress $f = \varepsilon E = \left[\frac{1}{16}/(12 \times 20)\right](29 \times 10^6) = 7550 \text{psi}$[③] }. Another very satisfactory method involves tightening the rods with some sort of sleeve nut or turnbuckle.

The preceding discussion on rods should illustrate why rolled shapes such as angles[④] have supplanted rods for most applications. In the early days of steel structures, tension members consisted of rods, bars, and perhaps cables. Today, although the use of cables is increasing for suspended-roof structures, tension members usually consist of single angles, double angles, tees, channels, W sections, or sections built up from plates or rolled shapes. These members look better than the old ones, are stiffer, and are easier to connect. Another type of tension section often used is the welded tension plate or flat bar, which is very satisfactory for use in transmission towers, signs, foot bridges, and similar structures.

The tension members of steel roof trusses may consist of single angles as small as $2\frac{1}{2} \times 2 \times \frac{1}{4}$ for minor members. A more satisfactory member is made from two angles placed back to back with sufficient space between them to permit the insertion of plates (called gusset plates) for connection purposes. Where steel sections are used back-to-back in this manner, they should be connected every 4 or 5 feet to prevent rattling, particularly in bridge trusses. Single angles and double angles are probably the most common types of tension members in use. Structural tees make very satisfactory chord members for welded trusses because web members can conveniently be connected to them.

For bridges and large roof trusses tension members may consist of channels, W or S shapes, or even sections built up from some combination of angles, channels, and plates. Single channels are frequently used, as they have little eccentricity and are conveniently connected. Although, for the same weight, W sections are stiffer than S sections. They may have a connection disadvantage in their varying depths. For instance. The $W12 \times 79$, $W12 \times 72$, and $W12 \times 65$ all have slightly different depths (12.38 inches, 12.25 inches, and 12.12 inches, respectively), while the S sections of a certain nominal size all have the same depths. For instance, the $S12 \times 50$, the $S12 \times 40.8$, and the $S12 \times 35$ all have 12.00 inches depths.

Although single structural shapes are a little more economical than built-up sections, the latter are occasionally used when the designer is unable to obtain sufficient area or rigidity from single shapes. Where built-up sections are used it is important to remember that field connections will have to be made and paint applied; therefore, sufficient space must be available to accomplish these things.

Members consisting of more than one section need to be tied together. Tie plates (also called tie bars) located at various intervals or perforated cover plates serve to hold the various pieces in their correct positions. These plates serve to correct any unequal distribution of loads between the various parts. They also keep the slenderness ratios of the individual parts within limitations and they may permit easier handling of the built-up members. Long individual members such as angles may be inconvenient to handle due to flexibility, but when four angles are laced together into one member, as shown in Fig. 15-1, the member has considerable stiffness. None of the intermittent tie plates may be considered to increase the effective areas of the sections. As they do not theoretically carry portions of the force in the main sections, their sizes are usually governed by specifications and perhaps by some judgment on the designer's part. Perforated cover plates are an exception to this rule, as part of their areas can be considered as being effective in resisting axial load.

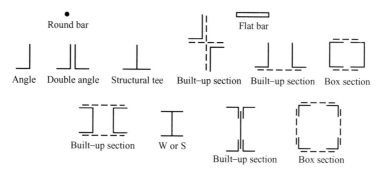

Fig. 15 – 1 **Types of tension members**

A few of the various types of tension members in general use are illustrated in Fig. 15 – 1. In this figure the dotted lines represent the intermittent tie plates or bars used to connect the shapes.

Steel cables are made with special steel alloy wire ropes which are cold-drawn to the desired diameter. The resulting wires with strengths of about 200000 to 250000 psi can be economically used for suspension bridges, cable supported roofs, ski lifts, and other similar applications. Normally, to select a cable tension member the designer uses a manufacturer's catalog. From the catalog the yield stress of the steel and the cable size required for the design force are determined. It is also possible to select clevises or other devices to use for connectors at the cable ends.

2. Axially loaded compression members

There are several types of compression members, the column being the best known. Among the other types are the top chords of trusses, bracing members, the compression flanges of rolled beams and built-up beam sections, and members that are subjected simultaneously to bending and compressive loads. Columns are usually thought of as being straight vertical members whose lengths are considerably greater than their thicknesses. Short vertical members subjected to compressive loads are often called struts or simply compression members.

There are three general modes by which axially loaded columns can fail. These are flexural buckling, local buckling and torsional buckling. These modes of buckling are briefly defined below.

(1) Flexural buckling (also called Euler buckling) is the primary type of buckling. Members are subject to flexure or bending when they become unstable.

(2) Local buckling occurs when some part or parts of the cross section of a column are so thin that they buckle locally in compression before the other modes of buckling can

occur. The susceptibility of a column to local buckling is measured by the width-thickness ratios of the parts of its cross section.

(3) Torsional buckling may occur in columns that have certain cross-sectional configurations. These columns fail by twisting (torsion) or by a combination of torsional and flexural buckling.

The longer a column becomes for the same cross section, the greater becomes its tendency to buckle and the smaller becomes the load it will support. The tendency of a member to buckle is usually measured by its slenderness ratio, which has previously been defined as the ratio of the length of the member to its least radius of gyration. The tendency to buckle is also affected by such factors as the types of end connections, eccentricity of load application, imperfection of column material, initial crookedness of columns, residual stresses from manufacture. etc.

The loads supported by a building column are applied by the column section above and by the connections of other members directly to the column. The ideal situation is for the loads to be applied uniformly across the column, with the center of gravity of the loads coinciding with the center of gravity of the column. Furthermore, it is desirable for the column to have no flaws, to consist of a homogeneous material, and to be perfectly straight, but these situations are obviously impossible to achieve.

Loads that are exactly centered over a column are referred to as axial or concentric loads. The dead loads may or may not be concentrically placed over an interior building column and the live loads may never be centered. For an outside column the load situation is probably even more eccentric, as the center of gravity of the loads will usually fall well on the inner side of the column. In other words, it is doubtful that a perfect axially loaded column will ever be encountered in practice.

The other desirable situations are also impossible to achieve because of the following: imperfections of cross-sectional dimensions, residual stresses, holes punched for bolts, erection stresses, and transverse loads. It is difficult to take into account all of these imperfections in a formula.

Slight imperfections in tension members and beams can be safely disregarded as they are of little consequence. On the other hand, slight defects in columns may be of major significance. A column that is slightly bent at the time it is put in place may have significant bending moments equal to the column load times the initial lateral deflection.

Obviously, a column is a more critical member in a structure than is a beam or tension member because minor imperfections in materials and dimensions mean a great deal. This fact can be illustrated by a bridge truss that has some of its members damaged by a truck. The bending of tension members probably will not be serious as the tensile loads will tend to straighten those members; but the bending of any compression members is a serious matter, as compressive loads will tend to magnify the bending in those members.

The preceding discussion should clearly show that column imperfections cause them to bend and the designer must consider stresses due to those moments as well as due to axial loads.

The spacing of columns in plan establishes what is called a bay. For instance, if the columns are 20 feet. on center in one direction and 25 feet in the other direction the bay size is 20 feet×25 feet. Larger bay sizes increase the user's flexibility in space planning. As to economy, a detailed study by John Ruddy indicates that when shallow spread footings are used, bays with length-to-width ratios of about 1.25 to 1.75 and areas of about 1000 square feet are the most cost efficient. When deep foundations are used, his study shows that larger bay areas are more economical.

3. Beams

1) Types of beams

Beams are usually said to be members that support transverse loads. They are probably thought of as being used in horizontal positions and subjected to gravity or vertical loads; but there are frequent exceptions—rafters, for example.

Among the many types of beams are joists, lintels, spandrels, stringers, and floor beams. Joists are the closely spaced beams supporting the floors and roofs of buildings, while lintels are the beams over openings in masonry walls such as windows and doors. A spandrel beam supports the exterior walls of buildings and perhaps part of the floor and hallway loads. The discovery that steel beams as a part of a structural frame could support masonry walls (together with the development of passenger elevators) is said to have permitted the construction of today's "skyscrapers". Stringers are the beams in bridge floors running parallel to the roadway, whereas floor beams are the larger beams in many bridge floors which are perpendicular to the roadway of the bridge and are used to transfer the floor loads from the stringers to the supporting girders or trusses. The term girder is rather loosely used but usually indicates a large beam and perhaps one into which smaller beams are framed.

2) Sections used as beams

The W shapes will normally prove to be the most economical beam sections and they have largely replaced channels and S sections for beam usage. Channels are sometimes used for beams subjected to light loads, such as purlins, and in places where clearances available require narrow flanges. They have very little resistance to lateral forces and need to be braced. The W shapes have more steel concentrated in their flanges than do S beams and thus have larger moments of inertia and resisting moments for the same weights. They are relatively wide and have appreciable lateral stiffness.

Another common type of beam section is the open web joist or bar joist. This type of section which is commonly used to support floor and roof slabs is actually a light shop-fabricated parallel chord truss. It is particularly economical for long spans and light loads.

Ⅰ. New Words

1. bay *n.* 开间,跨度
2. buckling *n.* 屈曲,压曲
3. stiffness *n.* 刚度
4. sag *vi.* &*n.* 下垂,松弛
5. fabricate *vt.* 制作,组合
6. sway *vi.* 摆动,侧移
7. sleeve *n.* 套筒,套管
8. nut *n.* 螺母,螺帽
9. turnbuckle *n.* 松紧螺丝扣,花篮螺栓,螺套,紧线器
10. gusset *n.* 节点板,连接板,角撑板
11. chord *n.* 弦,弦杆
12. slenderness *n.* 细长度
13. clevis *n.* U形钩,马蹄钩
14. gyration *n.* 旋转,回旋
15. crookedness *n.* 弯曲,扭曲,歪斜
16. rafter *n.* 椽,椽子
17. intel *n.* 过梁,楣
18. spandrel *n.* 拱肩,托梁
19. stringer *n.* 桁条,纵梁,纵枕木
20. purlin *n.* 檩,檩条

Ⅱ. Phrases and Expressions

1. bracing system   支撑体系,支撑系统
2. tie rod   拉杆,系杆
3. threaded rod   螺杆,螺纹杆
4. rule of thumb   经验法则
5. sleeve nut   套筒螺帽,套筒螺母
6. rolled shape   轧制钢材,轧制型材
7. slenderness ratio   长细比

Ⅲ. Notes

① "get one's feet wet" 是美国俚语,意思是 "to start learning something",译为 "着手,动手,亲自参加",有到实践中去学习的意思。

② "LRFD" 是 "Load and Resistance Factor Design" 的缩写,荷载抗力系数设计法。

③ "psi" 是计量单位(磅/平方英寸),英文全称为 "Pounds per square inch",1 psi= 6.895kPa。

④ "angle"是轧制型钢中的角钢。下文的"single angles，double angle"分别指单角钢、双角钢。

Ⅳ. Exercises

1. Translate the following words or phrases into Chinese.
（1）factored force
（2）sleeve nut
（3）flat bar
（4）rule of thumb
（5）buckling mode
（6）built-up section
（7）dead loads and concentric loads
（8）radius of gyration
（9）girder and truss
（10）open web joist

2. Translate the following sentences into English.
（1）钢棒通常刚度很小，且在自重下很容易弯曲。
（2）大跨度桁架中的拉杆可由角钢、槽钢或钢板等组合形成。
（3）钢索由冷拔到一定直径的特殊合金钢丝制成。
（4）柱子在平面上的间隔距离形成了所谓的开间。
（5）梁通常称为承受横向荷载的构件。

Ⅴ. Expanding

Know about the following terms related to steel members.
1. compressive strength 抗压强度
2. unconfined compressive strength 无侧限抗压强度
3. solid-web steel column 实腹式钢柱
4. built-up steel column; laced or battened compression member 格构式钢柱
5. composite steel and concrete beam 钢与混凝土组合梁
6. post-buckling strength of web plate 腹板屈曲后强度
7. overall stability 整体稳定

Ⅵ. Reading Material

Development of Steel Structures

1. Early uses of iron and steel

Although the first metal used by human beings was probably some type of copper alloy such as bronze (made with copper and tin and perhaps some other additives), the most important metal developments throughout history have occurred in the manufacture and use of iron and its famous alloy named steel. Today, iron and steel comprise almost 95 percent of all the tonnage of metal produced in the world.

Despite diligent efforts for many decades, archaeologists have been unable to discover when iron was first used. They did find an iron dagger and an iron bracelet in the Great Pyramid in Egypt, which they claim had been there undisturbed for at least 5000 years. The use of iron has had a great influence on the course of civilization since the earliest times, and may very well continue to do so in the centuries ahead. Since the beginning of the Iron Age in about 1000 B. C., the progress of civilization in peace and war has been heavily dependent on what people have been able to make with iron. On many occasions its use has decidedly affected the outcome of military engagements. For instance, in 490 B. C. in Greece at the Battle of Marathon, the greatly outnumbered Athenians killed 6400 Persians and lost only 192 of their own men. Each of the victors wore 57 pounds of iron armor in the battle. This victory supposedly saved Greek civilization for many years.

According to the classic theory concerning the first production of iron in the world, there was once a great forest fire on Mount Ida in Ancient Troy (now Turkey) near the Aegean Sea. The Land surface supposedly had a rich content of iron and the heat of the fire is said to have produced a rather crude form of iron which could be hammered into various shapes. Many historians believe, however, that human beings first learned to use iron that fell to the earth in the form of meteorites. Frequently the iron in meteorites is combined with nickel to produce a harder metal. Perhaps early human beings were able to hammer and chip this material into crude tools and weapons.

Steel is defined as a combination of iron with a small amount of carbon, usually less than 1 percent. It also contains small percentages of some other elements. Although some steel has been made for at least 2000 or 3000 years, there was really no economical production method available until the middle of the nineteenth century.

The first steel was surely obtained when the other elements necessary for producing it were accidentally present when iron was heated. As the years went by, steel probably was made by heating iron in contact with charcoal. The surface of the iron absorbed some carbon from the charcoal which was then hammered into the hot iron. Repeating this process several times resulted in a case-hardened exterior of steel. In this way the famous swords of Toledo and Damascus were produced.

The first large volume process for producing steel was named after Sir Henry Bessemer of England. He received an English patent of his process in 1855, but his efforts to obtain a U. S. patent for the process in 1856 were unsuccessful as it was proved that William Kelly of Eddyville, Kentucky, had made steel by the same process seven years before Bessemer applied for his English patent. Although Kelly was given the patent, the name Bessemer was used for the process.

Kelly and Bessemer learned that a blast of air through molten iron burned out most of the impurities in the metal. Unfortunately, at the same time the blow eliminated some desirable elements such as carbon and manganese. It was later learned that these needed elements could be restored by adding spiegeleisen, which is an alloy of iron, carbon, and manganese. It was further learned that the addition of limestone in the converter resulted in the removal of the phosphorus and most of the sulfur.

Before the Bessemer process was developed, steel was an expensive alloy used primarily for making knives, forks, spoons, and certain types of cutting toots. The Bessemer process reduced production costs greatly and allowed for the first time the production of large quantities of steel.

The Bessemer converter was commonly used in the United States until after the turn of the century, but since that time it has been replaced with better methods such as the open-hearth process and the basic oxygen process.

As a result of the Bessemer process, structural carbon steel could be produced in quantity by 1870, and by 1890 steel had become the principal structural metal used in the United States.

Today approximately 80 percent of the structural steel produced in the United States is made by melting scrap steel (primarily old car bodies) in electric furnaces. The molten steel is poured into molds which have approximately the final shapes of the members. The resulting sections are run through a series of rollers to squeeze them to their final shapes. The resulting members have better surfaces and fewer internal or residual stresses than newly made steel.

The term cast iron is used for the very low carbon content materials, while the very high carbon content materials are referred to as wrought iron. Steels fall in between cast iron and wrought iron and have carbon contents in the range of 0.15 percent to 1.7 percent.

The first use of metal for a sizable structure occurred in England in Shropshire (about 140 miles northwest of London) in 1779, when cast iron was used for the construction of the 100 ft. Coalbrookdale Arch Bridge over the River Severn. It is said that this bridge (which still stands) was a turning point in engineering history because it changed the course of the Industrial Revolution by introducing iron as a structural material. This iron was supposedly four times as strong as stone and thirty times as strong as wood.

A number of other cast-iron bridges were constructed in the following decades, but soon after 1840 the more malleable wrought iron began to replace cast iron. The development of the Bessemer process and subsequent advances such as the open-hearth process permitted the manufacture of steel at competitive prices which encouraged the beginning of the almost unbelievable developments of the last 100 years with structural steel.

2. Modern structural steels

The properties of steel can be greatly changed by varying the quantities of carbon present and by adding other elements such as silicon, nickel, manganese, and copper. A kind of steel that has a significant amount of the latter elements is referred to as an alloy steel. Although these elements do have a great effect on the properties of steel, the actual quantities of carbon or other alloying elements are quite small. For instance, the carbon content of steel is almost always less than 0.5 percent by weight and is normally from 0.2 to 0.3 percent.

The chemistry of steel is extremely important in its effect on such properties of the steel as weldability, corrosion resistance, resistance to brittle fracture, and so on. The ASTM specifies the exact maximum percentages of carbon, manganese, silicon, etc., which are permissible for a number of structural steels. Although the physical and mechanical properties of steel sections are primarily determined by their chemical composition, they are also influenced to a certain degree by the rolling process and by their stress history and heat treatment.

In the past few decades a structural carbon steel designated as A36 and having a minimum yield stress $F_y = 36$ ksi was the commonly used structural steel. However, most of the structural steel used in the United States is manufactured by melting scrap steel in electric furnaces. With this process the 50 ksi steel can be produced and sold at almost the same price as 36 ksi steel. As a result, at the time of this writing 50 ksi steel produced by the electric furnace process is the commonly used structural steel in the United States.

In recent decades the engineering and architecture professions have been continually requesting stronger steels, steels with more corrosion resistance, steels with better welding properties, and various other requirements. Research by the steel industry during this period has supplied several groups of new steels which satisfy many of the demands. Today there are quite a few structural steels designated by the ASTM and included in the LRFD Specification.

Structural steels are generally grouped into several major ASTM classifications: the all-purpose carbon steel (A36), the structural carbon steel (A529), the high-strength low-alloy structural steel (A572), the atmospheric corrosion-resistant high-strength low-alloy structural steels (A242 & A588) and quenched and tempered alloy steel plate (A514 & A852).

参 考 译 文

第15课 钢 构 件

1. 受拉构件

受拉构件常见于桥梁和屋面的桁架、塔架、支撑系统,以及其他用作拉杆的情况。用作受拉构件的截面的选择是设计中遇到的最简单的问题之一。由于没有屈曲的危险,设计者仅需要计算由构件承担的乘上分项系数的力的作用,并将该力除以设计应力(钢材强度设计值)以确定所需的有效横截面面积。然后,有必要去选择一个能提供所需面积的钢截面。虽然这些针对受拉构件介绍性的计算是相当简单的,但是它们确实满足了这样的重要任务,即让学生开始获得设计的理念,让他们涉足于大量的荷载抗力系数设计法手册。

受拉构件最简单的形式之一是圆形杆(圆钢),但是在将其连接到其他结构上时会存在某些困难。杆(圆钢)在过去常被使用,但是现在仅在支撑系统、轻型桁架和木结构中偶尔使用。杆(圆钢)不受设计师们欢迎的一个重要原因是过去常因使用不当而落下坏名;但是如果能正确设计和安装的话,它们在很多情况下是会满足要求的。

一般尺寸的杆(圆钢)几乎没有刚度,在其自重作用下很容易产生下垂,破坏了结构的外观。以前用于桥梁中的螺纹杆(有螺纹的圆钢),在工作时经常会松动和发出嘎嘎声。杆(圆钢)的另一个缺点是难以按要求的精确长度制作,随之而来的就是安装困难。

当杆(圆钢)被用作抗风支撑时,在杆(圆钢)中产生初始拉应力是一个好的方法,因为这将会拉紧结构,并减少嘎嘎声和摇摆。为了获得初始拉应力,构件可设计得比其所需长度稍短一些,这种方法对钢材制造者来说几乎没有什么麻烦。一个常用的经验法则是设计杆(圆钢)时每 20 英尺的长度缩短约 $\frac{1}{16}$ 英寸〔近似应力 $f=\varepsilon E=\left[\frac{1}{16}/(12\times 20)\right](29\times 10^6)=7550\mathrm{psi}$〕。另一种非常令人满意的方法包括采用某种套筒螺母或螺丝扣来拉紧杆(圆钢)。

前面关于杆(圆钢)的讨论应该阐明为什么诸如角钢的轧制型钢已经在多数应用中代替了杆(圆钢)。在早期的钢结构中,受拉构件是由杆(圆钢)、钢条,也可能是索等组成的。今天,尽管索在悬索屋顶结构中的应用不断增加,但是受拉构件通常包括单角钢、双角钢、T 形钢、槽钢、W 形钢或由钢板或轧制型钢组合而成的截面等。这些构件看起来比原来的更好、更结实、更容易连接。另一种常用的受拉构件是焊接的受拉钢板或扁钢,其非常适用于输电塔、广告牌、步行天桥及类似结构中。

对于小型构件而言,钢屋架的受拉构件可以由小至 $2\frac{1}{2}\times 2\times\frac{1}{4}$ 的单角钢组成。一种更令人满意的构件由背对背放置的双角钢制成,在它们之间具有足够的间距,以便于插入用于连接的板(称为节点板)。当钢截面采用这种背对背的方式时,应每隔 4 或 5 英尺连接一次,以防止发出嘎嘎声,特别是在桥梁桁架中。单角钢和双角钢可能是受拉构件中最常用的类型。对焊接桁架而言,结构 T 形钢可用作非常令人满意的弦杆,因为腹杆与它们连接起来很方便。

对于桥梁和大型屋架,受拉构件可以由槽钢、W 形钢或 S 形钢,甚至是由角钢、槽钢和钢板组合成的截面组成。单槽钢使用频繁,因为它们具有较小的偏心率并且易于连接。在重量相同的情况下,W 形钢比 S 形钢更结实,但是 W 形钢在厚度变化处会存在连接的缺陷。如 W 12×79,W 12×72 和 W 12×65 的厚度略有不同(分别为 12.38 英寸,12.25 英寸和 12.12 英寸),而具有一个特定公称尺寸的 S 形钢,其厚度相同,如 S 12×50,S 12×40.8 和 S 12×35 的厚度均是 12.00 英寸。

尽管单个结构型钢比组合截面更经济一些,但是当设计者不能从单个型钢获得足够的面积或刚度时,偶尔也会采用后者。使用组合截面时,重要的是要记住必须进行现场连接并涂漆;因此,要有足够的空间用以实施这些事情。

由多个截面钢组成的构件需要连接在一起。以不同间隔设置的连接板（也称为连接杆）或穿孔盖板用来将各个构件保持在其正确的位置上。这些板用于调整各个构件间的任何不相等的荷载分配。它们还将这些不同构件的长细比保持在限制范围之内，并且更易于处理组合构件（格构式构件）。长的单个构件（如角钢）可能因柔性大而不便于处理，但是当四个角钢组合成一个构件时（图15-1），该构件具有相当大的刚度。可以不考虑间断的连接板对截面的有效面积的增加。理论上它们不承担主截面中的部分力，因此它们的尺寸通常由规范或设计者方面的一些判断来确定。穿孔盖板是这一规则的例外，因为它们的部分截面被认为是可以有效抵抗轴向荷载的。

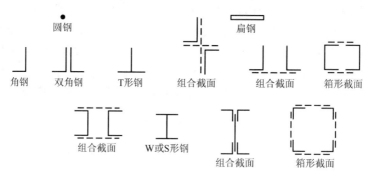

图15-1 受拉构件的类型

图15-1列举了各种常用受拉构件类型中的几种形式。在该图中，虚线表示用于连接型钢的间断的连接板或钢条。

钢索是由经冷拉至所需直径的特殊合金钢丝制成的。形成的强度约为200000～250000psi的钢丝，可经济地用于悬索桥、悬索屋顶、滑雪缆车（的索道）和其他类似应用。通常，为了选择索的受拉构件，设计者使用制造商的产品目录。从目录中，可以确定钢材的屈服应力和设计力所需要的缆索尺寸。也有可能选择U形夹具或其他装置用作索端的连接器。

2. 轴向受压构件

受压构件的类型很多，柱子是最为人所知的。在其他类型中，有桁架的上弦杆、支撑构件、轧制型钢梁和组合梁截面的受压翼缘，以及同时承受弯曲荷载和压力荷载的构件。柱子通常被认为是竖直构件，其长度远远大于其厚度。承受压力荷载的短的竖向构件经常被称为支柱或简单地称为受压构件。

轴向受压柱发生失效有三种常见模式。它们是弯曲屈曲、局部屈曲和扭转屈曲。这些屈曲模式简单地定义如下。

（1）弯曲屈曲（也称为欧拉屈曲）是屈曲的主要类型。当构件变得不稳定时，它们会挠曲或弯曲。

（2）局部屈曲的发生，是当柱子横截面的一部分或多个部分太薄，以至于它们在压力作用下，在其他屈曲模式发生之前就局部屈曲了。柱子对局部屈曲的敏感性可通过其横截面的宽厚比来衡量。

（3）扭转屈曲可能发生在具有某种横截面外形的柱子中。这些柱子是由于承受扭曲（扭转）或扭转和弯曲屈曲的组合作用而失效。

相同横截面的柱子越长，其屈曲的趋势就越大，其能承受的荷载也越小。构件屈曲的趋势通常根据其长细比来测定，它被定义为构件的长度与其最小回转半径之比。屈曲的趋势也受到诸如端部连接的类型、荷载的偏心、柱子材料的缺陷、柱子的初始弯曲、制造引起的残余应力等因素的影响。

建筑物的柱子承担的荷载是由上部的柱截面和由其他构件的连接直接施加到柱子上的。理想的情况是这些荷载均匀地被施加在柱子上，荷载的重心与柱子的重心一致。而且，理想的柱子是没有瑕疵，由均匀的材料组成，并且是完全笔直的，但是这些情况显然是不可能达到的。

准确地居中作用在柱子上的荷载被称为轴向或同轴荷载。恒荷载可以同心地或不同心地加载到建筑物的内柱上，但活荷载可能永远不会居中。对于外柱，荷载的作用位置甚至可能更加偏心，因为荷载的重心通常恰好落在柱子的内侧。换句话说，在实际中会遇到理想的轴向受荷的柱子是不太可能的。

由于下列原因，其他理想的情形也是不可能达到的：横截面尺寸的缺陷、残余应力、螺栓冲孔、装配应力和横向荷载等。在一个公式中很难考虑到所有的这些缺陷。

受拉构件和梁中的微小缺陷可以被安全地忽略掉，因为它们几乎没有什么影响。但是，柱子中微小的缺陷可能具有非常重要的意义。柱子在放置就位时，稍微的弯曲可能产生很大的弯矩，它等于柱子上的荷载乘以初始的横向挠度。

显然，与梁或受拉构件相比，柱子在结构中是较为关键的构件，因为材料和尺寸的微小缺陷都意味着很多。这个事实可以通过一个桥梁桁架来说明，该桥梁桁架有一些被卡车撞坏的构件。受拉构件的弯曲可能不是很严重的，因为拉伸荷载会趋向于把这些构件拉直；但是任何受压构件的弯曲都是一个严重的问题，因为压缩荷载会趋向于加大那些构件中的弯曲。

前面的讨论应该清楚地表明，柱子的缺陷导致它们弯曲，并且设计必须考虑由那些弯矩和由轴向荷载所引起的应力。

平面上柱的间距确定了什么是一个开间。例如，如果柱在一个方向的中心间距为20英尺，在另一个方向的中心间距为25英尺，则开间尺寸为20英尺×25英尺。较大的开间尺寸增加了使用者在空间布置中的灵活性。至于经济方面，约翰·拉迪的详细研究表明，当采用浅的扩展基础时，长宽比约为1.25～1.75、面积约为1000平方英尺的开间是最经济有效的。当采用深基础时，他的研究表明，较大的开间面积是更经济的。

3. 梁

1) 梁的类型

梁通常被认为是支撑横向荷载的构件。它们可能被认为是被用在水平位置并承受重力或垂直载荷；但是也常常有例外——如椽子。

在许多类型的梁中有搁栅、过梁、托梁、纵梁和楼面梁。搁栅是支撑建筑物的地板和屋顶的紧密布置的梁，而过梁是在砌体墙上开洞处如窗洞口和门洞口上面的梁。托梁支撑建筑物的外墙，以及可能支撑楼面和走廊的部分载荷。据说钢梁可作为结构框架的一部分能够支承砌体墙（加上载人电梯的发展）这一发现，使建造今天的"摩天大楼"成为可能。纵梁是在桥中平行于道路行进方向布置的梁，而楼面梁（横梁）是许多桥面中较大的梁，该梁垂直于桥的道路行进方向，被用于将楼面荷载从纵梁传递到支撑大梁或桁架上。术语"大梁"的采用相当地不严密,但通常指的是一根较大的梁,也可能是由较小的梁构成的大梁。

2)梁的截面

W 形钢通常被证明是最经济的梁截面,并且作为梁使用,它们已经大量地替代了槽钢和 S 形钢。槽形截面有时用于承受轻荷载的梁,如檩条,以及用在可利用的间隙要求狭窄翼缘的地方。它们对侧向力几乎没有抵抗力,因此需要被支撑。W 形钢具有比 S 形钢梁更多的集中在翼缘处的钢材,因此对于相同的重量,W 形钢梁具有较大的惯性矩和抵抗弯矩。它们相对较宽,并且具有明显的侧向刚度。

另一种常见类型的梁截面是空腹搁栅(托梁)或桁架搁栅。这种类型的截面通常用于支撑楼面和屋面板,它们实际上是一种由轻质的工厂制造的平行弦桁架。当大跨度和轻负荷时,它们是特别经济的。

Lesson 16

Steel Connections

For quite a few decades riveting was the accepted method used for connecting the members of steel structures. For the last few decades, however, bolting and welding have been the methods used for making structural steel connections, and riveting is almost extinct. This article is entirely devoted to bolted and welded connections.

1. Bolted connections

Bolting of steel structures is a very rapid field erection process that requires less skilled labor than does riveting or welding. This gives bolting a distinct economic advantage over the other connection methods in the United States where labor costs are so very high. Even though the purchase price of a high-strength bolt is several times that of a rivet, the overall cost of bolted construction is cheaper than that for riveted construction because of reduced labor and equipment costs and the smaller number of bolts required to resist the same loads.

There are several types of bolts that can be used for connecting steel members. They are described in the following paragraphs.

Unfinished bolts are also called ordinary or common bolts. They are classified by the ASTM[①] as A307[②] bolts and are made from carbon steels with stress-strain characteristics very similar to those of A36[③] steel. They are available in diameters from $\frac{5}{8}$ to $1\frac{1}{2}$ inches in $\frac{1}{8}$ inches increments.

A307 bolts generally have square heads and nuts to reduce costs, but hexagonal heads are sometimes used because they have a little more attractive appearance, are easier to turn and easier to hold with the wrenches, and require less turning space. As they have relatively large tolerances in shank and thread dimensions, their design strengths are appreciably smaller than those for high-strength bolts. They are primarily used in light structures subjected to static loads and for secondary members (such as purlins, girts, bracing, platforms, small trusses, and so forth).

Designers often are guilty of specifying high-strength bolts for connections when common bolts would be satisfactory. The strength and advantages of common bolts have usually been greatly underrated in the past. The analysis and design of A307 bolted connections are handled exactly as are riveted connections in every way except that the allowable stresses are slightly different.

High-strength bolts are made from medium carbon heat-treated steel and from alloy steel and have tensile strengths two or more times those of ordinary bolts. There are two basic types, the A325 bolts (made from a kind of heat-treated medium carbon steel) and the higher strength A490 bolts (also heat-treated but made from an alloy steel). High-strength bolts are used for all types of structures, from small buildings to "skyscrapers" and monumental bridges. These bolts were developed to overcome the weaknesses of rivets—primarily insufficient tension in their shanks after cooling. The resulting rivet tensions may not be large enough to hold them in place during the application of severe impactive and vibrating loads. The result is that they may become loose and vibrate and may eventually have to be replaced. High-strength bolts may be tightened until they have very high tensile stresses so that the connected parts are clamped tightly together between the bolt and nut heads, permitting loads to be transferred primarily by friction.

Sometimes high-strength bolts are made from A449 steel in sizes larger than the $1\frac{1}{2}$ inches maximum diameter A325 and A490 bolts. These larger bolts may also be used as high-strength anchor bolts and for threaded rods of many different diameters.

2. Welded connections

Welding is a process in which metallic parts are connected by heating their surfaces to a plastic or fluid state and allowing the parts to flow together and join (with or without the addition of other molten metal). It is impossible to determine when welding originated, but it was several thousand years ago. Metal-working, including welding, was quite an art in ancient Greece at least three thousand years ago, but welding had undoubtedly been performed for many centuries before those days. Ancient welding was probably a forging process in which the metals were heated to a certain temperature (not to the meeting stage) and hammered together.

Although modern welding has been available for a good many years, it has only come into its own in the last few decades for the building and bridge phases of structural engineering. The adoption of structural welding was quite slow for several decades because many engineers thought that welding had two great disadvantages that welds had reduced fatigue strength as compared with riveted and bolted connections and that it was impossible to ensure a high quality of welding without unreasonably extensive and costly inspection.

These negative feelings persisted for many years, although tests seemed to indicate that neither reason was valid. Regardless of the validity of these fears, they were widely held and undoubtedly slowed down the use of welding-particularly for highway bridges and, to an even greater extent, for railroad bridges. Today most engineers agree that welded joints have considerable fatigue strength. They will also admit that the rules governing the qualification of welders, the better techniques applied, and the excellent workmanship requirements of the AWS (American Welding Society) specifications make the inspection of welding a much less difficult problem. Furthermore, the chemistry of steels manufactured today is especially formulated to improve their weldability. Consequently, today welding is permitted for almost all structural work other than for some bridges.

On the subject of fear of welding, it is interesting to consider welded ships. Ships are subjected to severe impactive loadings that are difficult to predict, yet naval architects use all-welded ships with great success. A similar discussion can be made for airplanes and aeronautical engineers. The slowest adoption of structural welding was for railroad bridges. These bridges are undoubtedly subjected to heavier live loads than highway bridges, larger vibrations, and more stress reversals; but are their stress situations as serious and as difficult to predict as those for ships and planes?

Today it is possible to make use of the many advantages that welding offers since the fatigue and inspection fears have been largely eliminated. Several of the many welding advantages are discussed in the following paragraphs.

(1) To most persons the first advantage is in the area of economy, because the use of welding permits large savings in pounds of steel used. Welded structures allow the elimination of a large percentage of the gusset and splice plates necessary for bolted structures as well as the elimination of bolt heads. In some bridge trusses it may be possible to save up to 15 percent or more of the steel weight by using welding.

(2) Welding has a much wider range of application than bolting. Consider a steel pipe column and the difficulties of connecting it to other steel members by bolting. A bolted connection may be virtually impossible, but a welded connection will present few difficulties. The student can visualize many other similar situations in which welding has a decided advantage.

(3) Welded structures are more rigid because the members are often welded directly to each other. Frequently, the connections for bolted structures are made through intermediate connection angles or plates that deform due to load transfer, making the entire structure more flexible. On the other hand, greater rigidity can be a disadvantage where simple end connections with little moment resistance are desired. For such cases designers must be careful as to the type of joint they specify.

(4) The process of fusing pieces together gives the most truly continuous structures. It results in one-piece construction, and because welded joints are as strong as or stronger than the base metal, no restrictions have to be placed on the joints. This continuity advantage has permitted the erection of countless slender and graceful steel statically indeterminate frames throughout the world. Some of the more outspoken proponents of welding have referred to bolted structures, with their heavy plates and abundance of bolts, as looking like tanks or armored cars when compared with the clean, smooth lines of welded structures.

(5) It is easier to make changes in design and to correct errors during erection (and at less expense) if welding is used. A closely related advantage has certainly been illustrated in military engagements during the past few wars by the quick welding repairs made to

military equipment under battle conditions.

(6) Another item that is often important is the relative silence of welding. Imagine the importance of this fact when working near hospitals or schools or when making additions to existing buildings. Anyone will close-to-normal hearing who has attempted to work in an office within several hundred feet of a bolted job with attest to this advantage.

(7) Fewer pieces are used and, as a result, time is saved in detailing, fabrication, and field erection.

Ⅰ. New Words

1. rivet *n.* & *vt.* 铆钉
2. hexagonal *adj.* 六角形的，六边形的
3. wrench *n.* 扳手，扳钳，螺旋钳
4. girt *n.* 墙梁，围梁
5. monumental *adj.* 巨大的，不朽的
6. shank *n.* 无螺纹部，轴，杆
7. clamp *vt.*（用夹钳等）夹住，夹紧；*n.* 夹钳，夹具
8. forge *vt.* & *vi.* 锻造，铸造
9. weldability *n.* 可焊性
10. gusset *n.* 节点板，角撑板
11. fuse *vt.* & *vi.* 熔（化），熔合

Ⅱ. Phrases and Expressions

1. field erection　现场安装
2. light structure　轻型结构
3. unfinished bolt　粗制螺栓
4. high-strength bolt　高强度螺栓
5. static load　静荷载
6. live load　活荷载
7. impactive load　冲击荷载
8. vibrating load　振动荷载
9. fatigue strength　疲劳强度
10. splice plate　拼接板
11. statically indeterminate frame　超静定框架，超静定刚架
12. stress reversals　应力反复

Ⅲ. Notes

① "ASTM" 全称 "American Society for Testing and Materials"，是美国材料与试验协会的英文缩写。该协会是美国最老、最大的非营利性标准学术团体之一，工作中心是研

究和制定材料规范、试验方法标准，以及相关产品、系统、服务项目的特点和性能标准等。

② "A307"指按 ASTM 标准规定的普通螺栓性能等级，下文的 "A325" "A490" 指按 ASTM 标准规定的高强度螺栓性能等级。

③ "A36" 指按 ASTM 标准规定的钢材等级。

IV. Exercises

1. Translate the following phrases or sentences into Chinese.

(1) unfinished bolt

(2) secondary member

(3) monumental bridge

(4) anchor bolt

(5) threaded rod

(6) As they have relatively large tolerances in shank and thread dimensions, their design strengths are appreciably smaller than those for high-strength bolts.

(7) impactive load

(8) splice plate

(9) close-to-normal

(10) Welded joints are as strong as or stronger than the base metal.

2. Translate the following sentences into English.

(1) 螺栓连接与焊接广泛应用于钢结构。

(2) 高强度螺栓用热处理中碳钢以及合金钢制作。

(3) 焊接是将金属加热到液态，使其流动并连接在一起的过程。

(4) 与铆接和螺栓连接相比，焊接会降低连接的疲劳强度。

(5) 若采用焊接，在安装施工过程中会很容易修改设计和调整误差。

V. Expanding

Know about the following terms related to steel connection.

1. elective cross section area 有效横截面积
2. cover plate 盖板
3. covered electrode; covered electrode welding rod 焊条
4. welding wire 焊丝
5. welding flux; welding fluid 焊剂
6. quality grade of weld 焊缝质量级别
7. high-strength bolt with large hexagon head 大六角头型高强度螺栓
8. tor-shear type high-strength bolt 扭剪型高强度螺栓

Ⅵ. Reading Material

Portal Frames

Portal frames are generally low-rise structures, comprising columns and horizontal or pitched rafters, connected by moment-resisting connections. Resistance to lateral and vertical actions is provided by the rigidity of the connections and the bending stiffness of the members, which is increased by a suitable haunch or deepening of the rafter sections. This form of continuous frame structure is stable in its plane and provides a clear span that is unobstructed by bracing. Portal frames are very common, in fact 50% of constructional steel used in the UK is in portal frame construction. They are very efficient for enclosing large volumes, therefore they are often used for industrial, storage, retail and commercial applications as well as for agricultural purposes. This article describes the anatomy and various types of portal frame and key design considerations.

1. Anatomy of a typical portal frame

A portal frame building comprises a series of transverse frames braced longitudinally. The primary steelwork consists of columns and rafters, which form portal frames, and bracing. The end frame (gable frame) can be either a portal frame or a braced arrangement of columns and rafters (Fig. 16 – 1).

The light gauge secondary steelwork consists of side rails for walls and purlins for the roof. The secondary steelwork supports the building envelope, but also plays an important role in restraining the primary steelwork.

The roof and wall cladding separate the enclosed space from the external environment as well as providing thermal and acoustic insulation. The structural role of the cladding is to transfer loads to secondary steelwork and also to restrain the flange of the purlin or rail to which it is attached (Fig. 16 – 2).

2. Types of portal frames

Many different forms of portal frames may be constructed. Frame types described below give an overview of types of portal construction with typical features illustrated. This information only provides typical details and is not meant to dictate any limits on the use of any particular structural form.

1) Pitched roof symmetric portal frame (Fig. 16 – 3)

This kind of portal frame is generally fabricated from UB sections with a substantial eaves haunch section, which may be cut from a rolled section or fabricated from plate 25 to 35 meters are the most efficient spans.

2) Portal frame with internal mezzanine floor (Fig. 16 – 4)

Office accommodation is often provided within a portal frame structure using a partial width mezzanine floor.

Lesson 16 Steel Connections

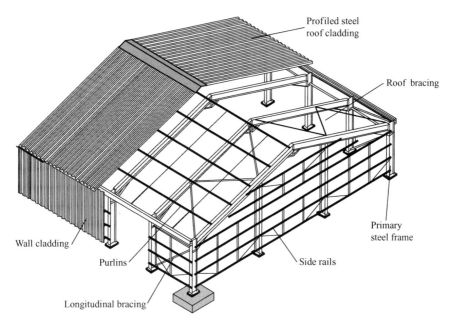

Fig. 16 – 1   Principal components of a portal framed building

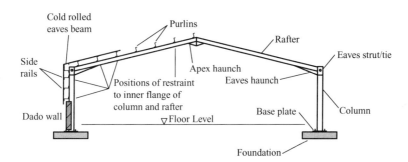

Fig. 16 – 2   Cross-section showing a portal frame and its restraints

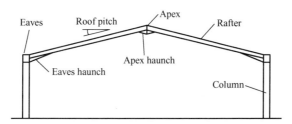

Fig. 16 – 3   Pitched roof symmetric portal frame

The assessment of frame stability must include the effect of the mezzanine; guidance is given in SCI P292.

3) Crane portal frame with column brackets (Fig. 16 – 5)

Where a travelling crane of relatively low capacity (up to say 20tonnes) is required, brackets can be fixed to the columns to support the crane rails. Use of a tie member or rigid column bases may be necessary to reduce the eaves deflection.

The spread of the frame at crane rail level may be of critical importance to the functioning of the crane; requirements should be agreed with the client and with the crane manufacturer.

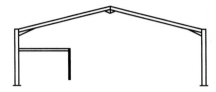 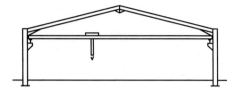

Fig. 16 – 4　Portal frame with internal mezzanine floor　　Fig. 16 – 5　Crane portal frame with column brackets

4) Tied portal frame (Fig. 16 – 6)

In a tied portal frame the horizontal movement of the eaves and the bending moments in the columns and rafters are reduced. A tie may be useful to limit spread in a crane-supporting structure.

The high axial forces introduced in the frame when a tie is used necessitate the use of second-order software when analyzing this form of frame.

5) Mono-pitch portal frame (Fig. 16 – 7)

A mono pitch portal frame is usually chosen for small spans or because of its proximity to other buildings. It is a simple variation of the pitched roof portal frame, and tends to be used for smaller buildings (up to 15-m span).

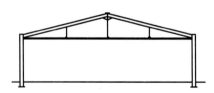 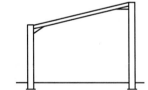

Fig. 16 – 6　Tied portal frame　　Fig. 16 – 7　Mono-pitch portal frame

6) Propped portal frame (Fig. 16 – 8)

Where the span of a portal frame is large and there is no requirement to provide a clear span, a propped portal frame can be used to reduce the rafter size and also the horizontal shear at the foundations.

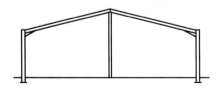

Fig. 16 – 8　Propped portal frame

7) Mansard portal frame (Fig. 16-9)

A mansard portal frame may be used where a large clear height at mid-span is required but the eaves height of the building has to be minimised.

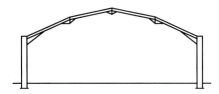

Fig. 16-9   Mansard portal frame

8) Curved rafter portal frame (Fig. 16-10)

Portal frames may be constructed using curved rafters, mainly for architectural reasons. Because of transport limitations rafters longer than 20 meters may require splices, which should be carefully detailed for architectural reasons.

The curved member is often modelled for analysis as a series of straight elements. Guidance on the stability of curved rafters in portal frames is given in SCI P281.

Alternatively, the rafter can be fabricated as a series of straight elements. It will be necessary to provide purlin cleats of varying height to achieve the curved external profile.

9) Cellular beam portal frame (Fig. 16-11)

Rafters may be fabricated from cellular beams for aesthetic reasons or when providing long spans. Where transport limitations impose requirement for splices, they should be carefully detailed, to preserve the architectural features.

The sections used cannot develop plastic hinges at a cross-section, so only elastic design is used.

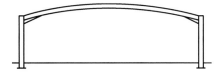

Fig. 16-10   Curved rafter portal frame

Fig. 16-11   Cellular beam portal frame

参 考 译 文

第 16 课　钢结构连接

在这几十年里，铆钉连接是钢结构中构件连接最常采用的方法。但是，在过去的几十年里，螺栓连接和焊缝连接已经成为钢结构连接的方式，而铆钉连接几乎被淘汰。本文主要讲解螺栓连接和焊缝连接。

1. 螺栓连接

钢结构的螺栓连接是一种非常快速的现场安装方式，与铆接或焊接相比，对工人的技术要求较低。在人工费用非常昂贵的美国，螺栓连接相对于其他连接方式具有明显的经济优势。尽管购买高强度螺栓的费用是购买铆钉费用的几倍，但是螺栓连接结构的整体费用比铆钉连接结构的费用低很多，这是由于减少了劳动力和设备的成本，以及抵抗同样荷载所需的螺栓数量较少。

有几种类型的螺栓可以用于连接钢结构构件。它们将在下面几个段落中进行描述。

粗制螺栓又被称为普通螺栓。美国试验材料学会将其归类为 A307 螺栓，而且此类螺栓是由碳素钢制成的，其应力应变特性与 A36 钢非常类似。这种螺栓的直径范围为 $\frac{5}{8} \sim 1\frac{1}{2}$ 英寸，且以 $\frac{1}{8}$ 英寸增加。

A307 螺栓通常采用方形头螺栓和螺母来降低成本；但是，有时也会使用六角头螺栓，这是由于六角头螺栓具有更吸引人的外观、更容易拧紧、更容易用扳手固定，而且所需的转动空间较小。由于这种螺栓在杆和螺纹尺寸上具有相对较大的偏差，所以其设计强度明显小于高强度螺栓的强度。通常情况下，这种螺栓用于承受静力荷载的轻型结构中，以及一些次要构件（如檩条、墙梁、支撑、平台以及小桁架等）。

设计者经常有这样的过失，就是普通螺栓就可满足的构件却通常指定高强度螺栓来连接。过去，普通螺栓的强度和优势都被很大程度地低估了。除了容许应力上的细微差别之外，A307 螺栓连接的分析和设计与铆钉连接的一样。

高强度螺栓是由热处理中碳钢或者合金钢制成，并且其拉伸强度是普通螺栓的两倍或更多倍。高强度螺栓有两种基本类型，分别是 A325 螺栓（由一种热处理中碳钢制成）和较高强度的 A490 螺栓（也是热处理钢，但由合金钢制成）。高强度螺栓可以用于各种类型的结构，从小型建筑物到"摩天大楼"，以及巨型桥梁。这些螺栓能够克服铆钉的缺点——主要是在冷却后其杆径的拉力（强度）不足。在连接处承受较强的冲击和振动荷载时，热处理后铆钉的拉力（强度）可能不够，无法将其位置固定。其结果是，铆钉连接变松并且振动，最终不得不将其更换。高强度螺栓应被拧紧使其具有较高的拉应力，这样，就能将螺栓和螺母头部之间的被连接件夹紧，荷载主要以摩擦进行传递。

有时，高强度螺栓由 A449 钢制成，而且其尺寸比最大直径为 $1\frac{1}{2}$ 英寸的 A325 和 A490 螺栓大。

2. 焊缝连接

焊接是指这样的一个过程，即通过把金属表面加热成塑态或流体状态让这些部分流在一起、熔在一起（添加或不添加其他熔融金属）的连接过程。我们无法确定焊接何时出现，但是可以确定的是至少是几千年前。包括焊接在内的金属加工工艺，在古希腊至少已有三千年的历史，但是毋庸置疑，焊接工艺比那时还要早几个世纪。古时的焊接很可能是一个将金属经过加热到一定的温度（但没有到达熔点），并锤打在一起的锻造过程。

尽管现代焊接已经应用很多年了，但是将其应用于结构工程的房屋和桥梁中仅仅只有几十年的时间。人们对结构焊接的接受过程经历了非常缓慢的几十年，主要是因为许多工程师认为，焊缝连接具有两大缺点：一是与铆钉连接和螺栓连接相比，焊接降低了疲劳强度；二是不经过大量的、昂贵的检验的情况下，不能确保高质量的焊接。

尽管试验表明这两个原因均是无根据的，但是这些负面的观点仍存续了很多年。不管这些担心的正确与否，这些担心仍被广泛接受并且毫无疑问地减缓了焊接的应用速度——尤其是在高速公路桥梁中，甚至是在铁路桥梁中的应用。现在，大多数的工程师都同意焊接节点具有相当大的疲劳强度。他们也将会承认，管理焊工资格的规则、更高的技术，以及 AWS（美国焊接协会）规定的较高的工艺要求等，使焊接检查不再是一个难题。此外，如今制造钢材的化学成分能调配以提高其可焊性。因此，现如今，除了用于桥梁之外，焊接几乎可以用于所有的结构工程中。

关于对焊接恐惧的问题，可考虑使用焊接船会很有趣。这些船需要承受很难预测的强大冲击荷载作用，而造船师们却只用焊接的方式成功地制造了船只。我们可以对飞机和航空工程师们进行相似的讨论。对结构焊接这种方式接受速度最慢的是在铁路桥梁中的应用。毫无疑问地，这些桥梁需要承受比高速公路桥梁更重的活荷载、更大的振动以及更多的应力反复作用；但是，它们的应力状态和船只与飞机的应力状态一样那么严重且难以预测吗？

如今，由于对疲劳以及检验的恐惧在很大程度上已经得到了消除，从而我们可以利用焊接的许多优势。在下面几个段落中来讨论焊接的几点优势。

（1）对大多数人而言，第一个优势就是经济性方面，因为焊接连接的使用可以大量地减少钢材的使用量。焊接结构能够减少用于螺栓连接结构中必需的大量节点板和拼接板，并能减少螺栓头的使用。在一些桥梁桁架中，通过使用焊缝连接，可能会节省15％或更多的钢材用量。

（2）与螺栓连接相比，焊缝连接具有更广泛的适用范围。细想一下钢管柱以及它与其他钢构件用螺栓连接的困难。用螺栓连接几乎是不可能的，而用焊缝连接的困难却非常少。学生可以在其他类似情形中发现焊接有许多明显的优势。

（3）焊接结构具有更大的刚性，这是因为各个构件彼此都直接焊接在一起。通常，在螺栓连接的结构中，连接是通过中间的连接角钢和节点板实现的，而这些连接件在传递荷载时会发生变形，这会使整个结构更加容易弯曲。另外，对于不需要抗弯承载力的简支端连接，较大的刚度就是一个缺点。针对这些情况，设计者们在选定连接形式时必须慎重。

（4）将各部分熔合在一起的过程得到了真正的连续结构。这样就形成一整块结构，而且由于焊接节点具有与被连接金属相同或更高的强度，所以在节点上不需设置任何约束。由这种连续性强的优势，使得在世界上建成了大量细长的薄柔型超静定钢框架结构。焊接的一些直接拥护者们指出，与具有简洁、平滑线条的焊接结构相比，带有厚重连接板和大量螺栓的螺栓结构看起来就像坦克或装甲车。

（5）如果使用焊缝连接，那么设计时做变更或施工时改正错误将会很容易（以较少的费用）。过去的几次战争中，在军队交战时，通过快速焊接修复可使军事装备一直处于作战状态，这就已经体现了一种紧密相关的优势。

（6）另一个重要的优点是焊接施工时相对安静。当在医院或学校附近施工时，或对已有建筑物进行扩建时，想象一下这一点的重要性。任何一个想要在距离采用螺栓连接工作点几百英尺的办公室内工作的、听力正常的人都能够证明此项优势。

（7）在焊接结构中采用的拼接板件很少，因此可以节省节点设计、制作加工以及现场安装的时间。

Lesson 17
Tall Buildings

The rapid growth of world civilization has a significant impact on the way humans live today. The conversion of agricultural land to development uses and the increasing urbanization of the world's population are making the building towards high vertically. More people go from urban to city, requiring more space for offices and for habitation; this increased the land use pressure and the average population density. The same area can not support sufficient facilities, the building need to be built taller and taller. The main factors which lead to the development of tall building are the following:

(1) Scarcity of land and spiraling rise in the cost of land;
(2) Increasing population and urbanization;
(3) Architectural requirements;
(4) Innovation of new structural system;
(5) Development of new material and technology.

Recent years, there have been immense development in the field of civil engineering in our country and they kept pace with rapid advances made in technology. One of which is the design and construction of tall buildings. A tall building is defined as one in which the structural system is modified to make it sufficiently economical to resist lateral forces due to wind or earthquakes, within the prescribe criteria for strength, drift and the comfort of occupants. ①

The early development of high-rise buildings began with structural steel framing. Reinforced concrete and stressed-skin tube systems have since been economically and competitively used in a number of structures for both residential and commercial purposes. The high-rise buildings ranging from 50 to 110 stories that are being built all over the United States are the result of innovations and development of new structural systems.

The vertical subsystems in a tall building transmit accumulated gravity load from story to story, thus requiring larger columns or wall sections to support such loading. ② In addition, these same vertical subsystems must transmit lateral loads, such as wind or seismic loads to the foundation. But more significantly, the overturning moment and the shear deflections produced by lateral forces are larger and must be carefully provided for design.

Tall buildings are constructed in the following forms:

Lesson 17 Tall Buildings

(1) Framed structure: Framed structures for resisting vertical and lateral loads have long been accepted as an important and standard means for designing buildings. They provide excellent opportunity for rectangular penetration of wall surfaces both within and at the outside of a building compared to shear wall structures. Framed systems are made up of beams and columns. The ability of tall building to resist the wind and other lateral forces depends on the rigidity of connections between the beams and columns, but the columns are made stronger when rigidly connected to resist the lateral as well as vertical forces through frame bending. They are adopted for low-and medium-rise buildings up to high-rise buildings, such as office, school, and residential use.

(2) Shear wall structure: When shear walls are compatible with other functional requirements, they can be economically utilized to resist lateral forces in high-rise buildings. They are more rigid, integrative than frame structure. The vertical and horizontal loads are resisted by the wall, but they can resist lateral load only in the plane of the walls (i. e. not in a direction perpendicular to them). Therefore, it is always necessary to provide shear walls in two perpendicular directions, or at least in sufficient orientation so that lateral force in any direction can be resist. In the past, the shear wall structure showed good seismic behavior and smaller damage. It is usefully employed for residential building.

(3) Frame-shear wall: It is composed of frame and shear wall. The vertical loads are resisted by frame and shear wall; and horizontal loads are resisted mainly by the shear wall. It is extensively used in high-rise office and hotel.

(4) Tube: <u>With the increase in height and story, the horizontal seismic action of tall building increases largely.</u> [3] Frame, shear wall and frame shear wall structures cannot satisfy this need, whereas tube has excellent wind and seismic resistance. The tube structure includes framed tube, tube in tube and bundled tube. The framed tube system consist of an interior shear wall tube resisting partially horizontal forces and an outer frame of spaced reasonably columns. A good example of this system is Jin Mao Building, Shanghai, 1999, 88 stories, Height: 421 meters (1381 feet). The tube-in-tube structural system (Fig. 17 - 1) is adopted for office building. It combines the interior shear wall tube with an outer framed tube. The bundled tube is also named combined tube. Several tubes combined in a plan to create large tube to make building more rigid, it is adopted for multifunctional high-rise building, for example, the Sears tower (Fig. 17 - 2), Chicago, United States which was completed in 1974, 442 meters (1450 feet).

(5) Towers: Tall structures with relatively small cross-section and with a large ratio between the height and maximum width are known as towers. We often see water towers, radio and television towers, and transmission line towers. The famous Eiffel Tower, in Paris, France, 300-m high, is constructed in 1889. It is the first structure to exceed 300 meters in height. The addition of a telecommunications tower in the 1950s brought the overall height to 324 meters.

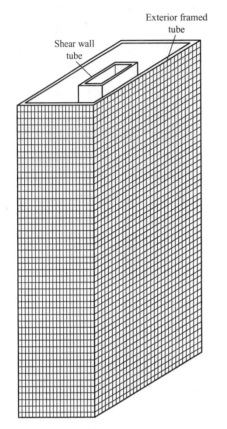

Fig. 17-1 Schematic sketch of tube-in-tube system

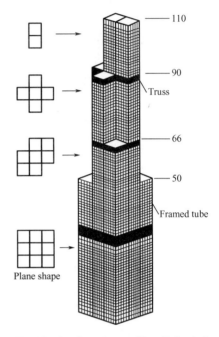

Fig. 17-2 Sears tower (bundled tube)

(6) Silos: It is defined as large size containers which are used to store grains, cement, coal, etc. In general, the shapes of silos are of circular cross-section.

In order to make tall building highly efficiently, designers need to select different structural systems according to projects. Meanwhile, they should consider various criteria, such as load, strength, stability, durability, stiffness and drift, foundation settlement, creep, shrinkage and temperature effects, fire, human comfort criteria.

Ⅰ. New Words

1. agricultural *adj.* 农业的，农艺的
2. facility *n.* 设施，设备
3. spiral *n.* 螺旋，螺纹的一圈
4. innovation *n.* 革新，改革，创新
5. lateral *adj.* 侧面的，横向的
6. drift *n.* 侧移，偏移
7. accumulate *vt.* 积累，累积
8. seismic *adj.* 地震的，因地震而引起的
9. connection *n.* 连接，连接件

10. bundle *n.* 束，捆
11. tower *n.* 塔，塔楼
12. telecommunication *n.* 电信，远程通信，无线电通信
13. container *n.* 容器

Ⅱ. Phrases and Expressions

1. high-rise building　高层建筑
2. defined as　被定义为
3. stressed-skin tube　薄壳筒体
4. turning moment　倾覆力矩
5. shear deflections　剪切变形
6. tube-in-tube　筒中筒
7. rigidity of connection　刚性连接
8. seismic behavior　抗震性能
9. foundation settlement　基础沉降

Ⅲ. Notes

① 此句中的"one"和"it"在本文中均指"tall building"，译为"高层建筑"。"in which"指"在高层建筑中"。

② "subsystem"的原意为"子系统"，"sub-"在英语中作为前缀，常译为"低级的，从属的"。

③ "with the increase in height and story"在句子中作伴随状语，译为"随着高度和层数的增加"，此句的真正主语为"the horizontal seismic action of tall building"，谓语动词为"increase"，状语为"largely"，表示程度。

Ⅳ. Exercises

Fill in the blanks with the information given in the text.

1. In addition, these same _____ subsystems must transmit _____ loads, such as _____ or _____ loads to the foundation.

2. When shear walls are compatible with other functional requirements, they can be _____ utilized to resist _____ forces in high-rise buildings. They are more _____, _____ than frame structure.

3. It is composed of _____ and _____. The vertical loads are resisted by _____ and _____; and horizontal loads are resisted mainly by the _____.

Ⅴ. Expanding

Know about the following terms related to tall building.

1. ultrahigh-rise building   超高层建筑
2. column-diagonal truss tube   柱列对角桁架筒体
3. ductile frame   延性框架
4. mega frame   巨型框架
5. giant space truss   巨型空间桁架
6. transformation layer   转换层
7. antidumping moment   抗倾覆力矩
8. storey drift   层间位移
9. inter-story elastic displacement   弹性层间位移

Ⅵ. Reading Material

The Future of Tall Building

Zoning effects on the density of tall building and solar design may raise ethical questions. Energy limitations will continue to be a unique design challenge. A combined project of old and new buildings may bring back human scale to our cities. Owners and conceptual designers will be challenged in the 1980s to produce economically sound, people-oriented buildings.

In 1980 the Lever House, designed by Skidmore, Owings and Merrill (SOM) received the 25-year award from the American Institute of Architects "in recognition of architectural design of enduring significance". This award is given once a year for a building between 25 and 35 years old. Lewis Mumford described the Lever House as "the first office building in which modern materials, modern construction, modern functions have been combined with a modern plan". At the time, this daring concept could only be achieved by visionary men like Gordon Bunshaft, the designer, and Charles Luckman, the owner and then-president of Lever Brothers. The project also included a few "first": (1) it was the first sealed glass tower ever built; (2) it was the first office building designed by SOM; and (3) it was the first office building on Park Avenue to omit retail space on the first floor. Today, after hundreds of look-alikes and variations on the grid design, we have reached what may be the epitome of tall building design: the nondescript building. Except for a few recently completed buildings that seem to be people-oriented in their lower floors, most tall buildings seem to be a repetition of the dull, graph-paper-like monoliths in many of our cities. Can this be the end of the design-line for tall building? Probably not. There are definite signs that are most encouraging. Architects and owners have recently begun to discuss the design problem publicly. Perhaps we are at the threshold of a new era. The 1980s may bring forth some new visionaries like Bunshaft and Luckman. If so, what kinds of restrictions or challenges do they face?

1. Zoning

Indications are strong that cities may restrict the density of tall buildings, that is, reduce the number of tall buildings per square mile. In 1980 the term grid-lock was used for the first time publicly in New York City. It caused a terror-like sensation in the pit of one's stomach. The term refers to a situation in which traffic comes to a standstill for many city blocks in all directions. The jam-up may even reach to the tunnels and bridges. Strangely enough, such an event happened in New York in a year of fuel shortages and high gasoline prices. If we are to avoid similar occurrences, it is obvious that the density of people, places, and vehicles must be drastically reduced. Zoning may be the only long-term solution.

Solar zoning may become more and more popular as city residents are blocked from the sun by tall buildings. Regardless of how effectively a tall building is designed to conserve energy, it may at the same time deprive a resident or neighbor of solar access. In the 1980s the right to see the sun may become a most interesting ethical question that may revolutionize the architectural fabric of the city. Mixed-use zoning which become a financially viable alternative during the 1970s, may become commonplace during the 1980s, especially if it is combined with solar zoning to provide access to the sun for all occupants.

2. Renovation

Emery Roth and Sons designed the Palace Hotel in New York as an addition to a renovated historic Villard house on Madison Avenue. It is a striking example of what can be done with salvageable and beautifully detailed old buildings. Recycling both large and small buildings may become the way in which humanism and warmth will be returned to buildings during the 80's. If we must continue to design with glass and aluminum in stark grid patterns, for whatever reason, we may find that a combination of new and old will become the great humane design trend of the future.

3. Conceptual design

It has been suggested in architectural magazines that the Bank of America office building in San Francisco is too large for the city's scale. It has also been suggested that the John Hancock Center in Boston is not only out of scale but also out of character with the city. Similar statements and opinions have been made about other significant tall buildings in cities throughout the world. These comments raise some basic questions about the design process and who really makes the design decisions on important structures — and about who will make these decisions in the 1980s.

Will the forthcoming visionaries — architects and owners — return to more humane designs?

Will the sociologist or psychologist play a more important role in the years ahead to help convince these visionaries that a new, radically different, human-scaled architecture is long overdue? If these are valid questions, could it be that our best architectural designers of the 60's and 70's will become the worst designers of the 80's and 90's? Or will they learn and respond to a valuable lesson they should have learned in their "History of

Architecture" course in college that "architecture usually reflects the success or failure of a civilized society"? Only time will tell.

参 考 译 文

第 17 课　高 层 建 筑

　　世界文明的快速发展对当今人类生活方式产生了重大影响。农业用地向开发利用的转变和世界人口都市化的增长使得建筑物向高发展。更多的人从农村走向城市，需要更多的空间用于办公和居住，这就会增加土地使用的压力和人口的平均密度。相同的区域不能提供足够的设施，建筑物就需要建得越来越高。导致高层建筑发展的主要因素如下：

　　（1）土地的缺乏以及地价的螺旋式上升；

　　（2）人口的增长和都市化；

　　（3）建筑需要；

　　（4）新型结构体系的革新；

　　（5）新材料和技术的发展。

　　最近几年，我国土木工程领域取得了巨大发展，并与技术的快速进步保持同步，其中之一是高层建筑设计和施工（的快速发展）。高层建筑可解释为改进结构体系使其足够经济以抵抗由风和地震引起的侧向力，以满足规定的强度、侧移和居住的舒适性标准。

　　早期的高层建筑发展始于钢框架结构，后来用于居住和商业用途的许多建筑采用了既经济又有竞争性的钢筋混凝土体系和外壳承载式筒体体系。正在美国各地修建的 50～110 层的高层建筑正是这些新的结构体系创新和发展的结果。

　　高层建筑的竖向结构体系从上到下逐层对累积的重力荷载进行传递，这就需要有较大尺寸的柱或墙体截面来承担荷载。另外，这些竖向结构体系还要将风荷载及地震荷载等侧向荷载传给基础。但是更重要的是侧向力产生的倾覆力矩和剪切变形要大得多，必须谨慎设计才能保证。

　　高层建筑的形式有如下几种。

　　（1）框架结构：用于抵抗竖向和水平荷载的框架结构，长期以来被公认为是一种设计建筑物的重要的和标准的形式。与剪力墙结构相比，这种结构更适合在建筑物内部或外墙上开设矩形洞口。框架结构体系是由梁、柱组成的。高层建筑抵抗风荷载和其他侧向力的能力取决于梁和柱之间连接的刚度，但当柱与梁刚性连接时，是通过框架受弯来抵抗水平和竖向荷载，因此柱要做得强些。这种体系适用于低层、多层及高层建筑，如办公楼、学校和住宅。

　　（2）剪力墙结构：当剪力墙能满足其他功能要求时，高层建筑中采用剪力墙来抵抗侧向荷载更经济。剪力墙结构比框架结构的刚度和整体性更好。竖向荷载和水平荷载由墙体来抵抗，但是，剪力墙只能抵抗平行于墙平面的荷载（也就是说不能抵抗垂直于墙面的荷载）。因此，总是需要在相互垂直的两个方向提供剪力墙，或至少在有效方向提供剪力墙以抵抗任何方向的水平力。过去，剪力墙结构显示出良好的抗震性能和较小的破坏。这种体系常用于居住建筑。

　　（3）框架-剪力墙结构：它是由框架和剪力墙组成的，垂直荷载由框架和剪力墙承担，水平荷载主要由剪力墙承担。这种结构广泛应用于高层办公楼和宾馆。

（4）筒体结构：随着高度和层数的增加，高层建筑的水平地震作用极大地增加，框架、剪力墙和框架-剪力墙结构已不能满足需求，而筒体具有良好的抗风和抗震性能。筒体结构包括框筒、筒中筒和束筒。框筒体系由抵抗部分水平力的内部剪力墙核心筒和由间距合理的柱组成的外部框架组成。上海金茂大厦，建于1999年，88层，高421m（1381英尺）就是一个很好的案例。筒中筒体系（图17-1）适用于办公建筑，它是由内部的剪力墙核心筒与外部的框筒组成。束筒体系也称为组合筒，几个筒在平面上组成一个大筒使建筑物刚度更大，它适用于多功能高层建筑，如美国芝加哥的西尔斯大厦（图17-2），1974年完工，高442m（1450英尺）。

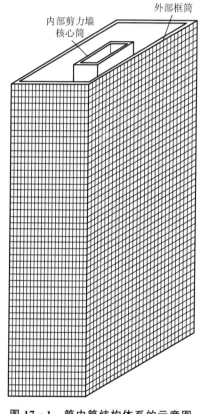

图17-1 筒中筒结构体系的示意图

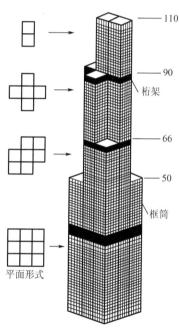

图17-2 西尔斯大厦（束筒体系）

（5）塔：具有截面积小、高宽比大特点的高层建筑被称之为塔。我们经常看到水塔、广播电视塔以及传输线塔。法国巴黎著名的埃菲尔铁塔，高300m，建于1889年，它是第一个高度超过300m的建筑。在20世纪50年代又增添了一座无线电通信塔，总高度达到324m。

（6）筒仓：可以定义为用来储存粮食、水泥、煤等的一个大尺寸容器。一般地，筒仓的形状都是圆形截面。

为了使高层建筑效能更高，设计师需要根据项目选择不同的结构体系。同时，他们还要考虑不同的准则，如荷载、强度、稳定性、耐久性、刚度和侧移，基础沉降、徐变、收缩和温度影响、火，以及人的舒适性。

Lesson 18

Prestressed Concrete

Concrete is strong in compression, but weak in tension: its tensile strength varies from 8 to 14 percent of its compressive strength. Due to such a low tensile capacity, flexural cracks develop at early stages of loading. In order to reduce or prevent such cracks from developing, a concentric or eccentric force is imposed in the longitudinal direction of the structural element. This force prevents the cracks from developing by eliminating or considerably reducing the tensile stresses at the critical midspan and support sections at service load, thereby raising the bending, shear, and torsional capacities of the sections. The sections are then able to behave elastically, and almost the full capacity of the concrete in compression can be efficiently utilized across the entire depth of the concrete sections when all loads act on the structure.

The development of early cracks in reinforced concrete due to non-compatibility in the strains of steel and concrete was perhaps the starting point for the development of a new material like "prestressed concrete".

Prestressed concrete is not a new concept, dating back to 1872, when P. H. Jackson, an engineer from California, patented a prestressing system that used a tie rod to construct beams or arches from individual blocks. After a long lapse of time during which little progress was made because of the unavailability of high-strength steel to overcome prestress losses, R. E. Dill of Alexandria, Nebraska, recognized the effect of the shrinkage and creep (transverse material flow) of concrete on the loss of prestress. In the early 1920s, W. H. Hewett of Minneapolis developed the principles of circular prestressing.

Eugene Freyssinet proposed methods to overcome prestress losses through the use of high-strength and high-ductility steels in 1926 — 1928. In 1940, he introduced the new well-known and well-accepted Freyssinet system.

Prestressed concrete is an improved form of reinforcement. Steel rods are bent into the shapes to give them the necessary degree of tensile strength.[①] They are then used to prestress concrete, usually by one of two different methods. The first is to leave channels in a concrete beam that correspond to the shapes of the steel rods. When the rods are run through the channels, they are then bonded to the concrete by filling the channels with grout, a thin mortar of binding agent. In the other (and more common) method, the prestressed steel rods are placed in the lower part of a form that corresponds to the shape of the finished structure, and the concrete is poured around them. Two methods are referred to as "pre-tensioned method" and "post-tensioned method". Because prestressed concrete is so economical, it is a highly desirable material.

From the preceding discussion, it is plain that permanent stresses in the prestressed structural member are created before the full dead and live loads are applied in order to eliminate or considerably reduce the net tensile stresses. With reinforced concrete, it is assumed that the tensile strength of the concrete is negligible and disregarded. This is because the tensile forces resulting from the bending moments are resisted by the bond created in the reinforcement process. Cracking and deflection are therefore essentially irrecoverable in reinforced concrete once the member has reached its limit state at service load.

The reinforcement in the reinforced concrete member does not exert any force of its own on the member, contrary to the action of prestressing steel. The steel required to produce the prestressing force in the prestressed member actively preloads the member, permitting a relatively high controlled recovery of cracking and deflection. Once the flexural tensile strength of the concrete is exceeded, the prestressed member starts to act like a reinforced concrete element.

Two types of bond stress must be considered in the case of prestressed concrete. The first of these is referred to as "transfer bond stress" and has the function of transferring the force in a pre-tensioned tendon to the concrete. The second type of bond is termed "flexural bond stress" and comes into existence in pre-tensioned and bonded, post-tensioned members when the members are subjected to external loads.

Bond stresses also occur between the tendons and the concrete in both pre-tensioned and bonded, post-tensioned members, as a result of changes in the external load. There are of course no transfer bond stresses in post-tensioned members, since the end anchorage device relatively low in prestressed members for loads less than the cracking load, there is an abrupt and significant increase in these bond stresses after the cracking load is exceeded. Because of the indeterminancy which results from the plasticity of the concrete for loads exceeding the cracking load, accurate computation of the flexural-bond stresses cannot be made under such conditions. Again, tests must be relied upon as a guide for design.

Prestressed concrete uses less steel and less concrete. Due to the utilization of concrete in the tension zone, a saving of 15 to 30 percent in concrete is possible in comparison with reinforced concrete. The savings in steel are even higher, 60 to 80 percent, mainly due to the high permissible stresses allowed in the high tensile wires. Although there is considerable saving in the quantity of materials used in prestressed concrete members in comparison with reinforced concrete members, the economy in cost is not that significant due to the additional costs incurred for the high strength concrete high tensile steel, anchorages, and other hardware required for the production of prestressed members. In spite of these additional costs, if a large enough number of precast units are manufactured. The difference between at least the initial costs of prestressed and reinforced concrete systems is usually not very large. And the indirect long-term savings are quite substantial, because less maintenance is needed, a longer working life is possible due to better quality control of the concrete, and lighter foundations are achieved due to the smaller cumulative weight of the superstructure.

The economy of prestressed concrete is also well established for long-span structures. According to Dean, standardized precast bridge beams between 10 and 30 meters long and precast prestressed piles have proved to be economical than steel and reinforced concrete in the United States. According to Abeles, precast prestressed concrete is economical for floors, roofs and bridges of spans up to 30 meters and for cast in situ work, it applies to spans up to 100 meters. In the long span range, prestressed concrete is generally economical in comparison with reinforced concrete and steel construction.

Prestressed concrete offers great technical advantages in comparison with other forms of construction, such as reinforced concrete and steel. In the case of fully prestressed members, free from tensile stresses under working loads, the cross-section is more efficiently utilized when compared with a reinforced concrete section which is cracked under working loads. Within certain limits, a permanent dead load may be counteracted by increasing the eccentricity of the prestressing force in a prestressed structural element, thus effecting saving in the use of materials.

<u>A prestressed concrete flexural member is stiffer under working loads than a reinforced concrete member of the same depth.</u>② However, after the onset of cracking, the flexural behavior of a prestressed member is similar to that of a reinforced concrete member. Prestressed concrete members possess improved resistance to shearing forces, due to the effect of compressive prestress, which reduces principal tensile stress. The use of curved cables, particularly in long-span members helps to reduce the shear forces developed at the support sections.

<u>The use of high strength concrete and steel in prestressed members results in lighter and slender members than could be possible by using reinforced concrete.</u>③ The two structural features of prestressed concrete, namely high strength concrete and freedom from cracks, contributes to the improved durability of the structure under aggressive environmental conditions. Prestressing of concrete improves the ability of the material for energy absorption under impact loads. The ability to resist repeated working loads has been proved to be as good in prestressed as in reinforced concrete.

Prestressed concrete has made it possible to develop buildings with unusual shapes, like some of the modern sports arenas, with large spaces unbroken by any obstructing supports. The uses for this relatively new structural method are constantly being developed.

Today, prestressed concrete is used in buildings, underground structures, TV towers, floating storage and offshore structures, power stations, nuclear reactor vessels, and numerous types of bridge systems including segmental and cable-stayed bridges. They demonstrate the versatility of the prestressing concept and its all-encompassing application.

The success in the development and construction of all these structures has been due in no small measures to the advances in the technology of materials, particularly prestressing steel, and the accumulated knowledge in estimating the short-term and long-term losses in the prestressing forces.

Ⅰ. New Words

1. prestress *n.* 预应力；*vt.* 给……预加应力
2. crack *n.* 裂缝
3. concentric *adj.* 同轴的，同中心的
4. eccentric *adj.* 偏心的，不同轴的
5. midspan *n.* 跨中
6. pretension *n.* 张拉
7. anchorage *n.* 锚具
8. indeterminancy *n.* 不确定性
9. eccentricity *n.* 偏心距
10. segmental *adj.* 分节的，分段的
11. versatility *adj.* 多功能性的

Ⅱ. Phrases and Expressions

1. loss of prestress   预应力损失
2. pre-tensioned method   先张法
3. post-tensioned method   后张法
4. ensile stress   拉应力
5. bond stress   黏结应力
6. cracking load   开裂荷载
7. tension zone   受拉区
8. flexural member   受弯构件
9. long span structures   大跨度结构

Ⅲ. Notes

① "prestressed" 一词的前缀 "pre-" 表示 "……前的，预先"，如 precast（预制），prefabricate（预制），premix（预先拌和），pretension（预张拉）等。

② "is stiffer under working loads than" 是比较级的用法，译为 "比……具有更大的刚度"；"under working loads" 作为插入语，不影响句子结构，译为 "在工作荷载下"。

③ "could be possible by using reinforced concrete" 为省略句，省略主语 "it"，"it could be possible by using reinforced concrete members" 译为 "使用钢筋混凝土构件是可能的"。

Ⅳ. Exercises

Fill in the blanks with the information given in the text.

1. In order to reduce or prevent such cracks _____ developing, a force is imposed in the longitudinal direction _____ the structural element.

2. The development of early cracks in reinforced concrete non-compatibility in the strains of _____ and _____ was perhaps the starting point for the development of a new material like.

3. _____ concrete offers great technical advantages other forms of construction, such as reinforced concrete and steel.

Ⅴ. Expanding

Remember the following terms related to the prestressed concrete.

1. prestress degree   预应力度
2. tensioning end   张拉端
3. fixed end   固定端
4. tension control stress   张拉控制应力
5. prestress value   预应力值
6. crack resistance   抗裂度
7. prestressed steel   预应力钢筋
8. precompression zone   预压区
9. pretension area   预拉区
10. unbonded prestressing   无黏结预应力

Ⅵ. Reading Material

Method of Prestressing the Concrete

The pre-compressing of the tension zone concrete with a hydraulic jack as illustrated in the example is feasible. Indeed, the first successfulprestressed concrete structure, an arch bridge designed by the eminent French engineer E. Freyssinet in 1928, is prestressed with hydraulic jacks pressing the arch ring. However, the most extensively used method of prestressing the concrete is to compress the concrete with stretched steel, which is termed the tendon. Two basic procedures have been developed.

1. The pre-tension method

In the pre-tension method, the tendons are first stretched and anchored on abutments at both ends [Fig. 18 – 1(a)]. Then the concrete member is cast around the tendon [Fig. 18 – 1(b)]. After the concrete has acquired prescribed strength, the tendons are released from the abutments [Fig. 18 – 1(c)], the stretching force of the tendon is transmitted to the concrete by bond and the concrete is compressed.

Lesson 18 Prestressed Concrete

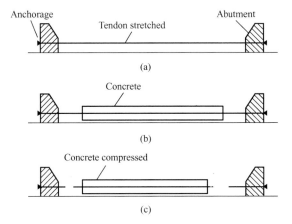

Fig. 18 – 1 Pre-tension method

2. The post-tension method

In the post-tension method, the member is first cast with a duct in it for the tendons [Fig. 18 – 2(a)]. After the concrete has attained a prescribed strength, the tendons are threaded through the duct and anchored at one end on the member with an anchoring device. Then the tendons are stretched at the other end with a jack using the member as a bed [Fig. 18 – 2(b)]. Thus the stretching of the tendon takes place simultaneously with the compressing of concrete. After the stretching end is also anchored on the member [Fig. 18 – 2(c)], the procedure is finished with the duct grouted or not grouted with cement concrete. The stretching force of the tendon is transmitted to the concrete through the anchoring devices at both ends of the tendon.

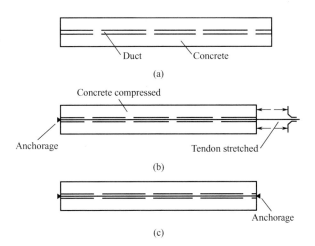

Fig. 18 – 2 Post-tension method

3. Pre-tension vs. post-tension

The pre-tension method is simple and economic. However, the abutments are massive permanent installment, and the prestressing is usually done in a casting yard of a prefabrication plant. As there are the problems of the transportation of the prestressed member to the job-site and the hoisting of it in the assembling of the structure, the pre-tension method is more adaptable for small sized-members. The post-tension method is more elaborate in its process, and the anchoring devices permanently attached on the member are expensive hardware. But the necessary equipment for stretching the tendons is portable and can be used in the field. Therefore the post-tension method is more adaptable for large sized members.

In the pre-tension method, the tendons are stretched between the abutments without intermediate support, so the profile of the tendons must run straight. In the post-tension method, the profile of the tendons can follow the profile of the duct in the cast member, and can assume the most optimum shape for the intended load.

The pre-tension method is so termed because the tendons are stretched before the concrete is cast. The post-tension method is so termed because the tendons are stretched after the concrete is cast. However, the fundamental difference between these two methods lies in the way the compression is transmitted to the concrete. In the pre-tension method, the compression is transmitted to the concrete through the bond between the tendons and the concrete. In the post-tension method, the compression is transmitted to the concrete through the anchoring devices at the ends of the tendons. There are numerous other ways to prestress a concrete structure, but they can always be classified into these two categories of pre-tension and post-tension by this fundamental difference. For instance, expansion cement may be used in the mix so that when the concrete expands during setting, it is compressed by the restriction of the reinforcement. Clearly this procedure belongs to the category of pre-tension. Steel wires in a high strung state may be wound on the outside of a concrete liquid container or a concrete pipe to exert compression on the concrete so that it may not leak under internal pressure. This is a case belongs to the category of post-tension.

参 考 译 文

第 18 课 预应力混凝土

混凝土的抗压强度很高,但抗拉强度却很低:其抗拉强度是抗压强度的 8%～14%。由于这种较低的抗拉能力,在加载的早期阶段,就容易产生弯曲裂缝。为了减少或阻止这些裂缝的发展,可在结构构件的纵向施加一个轴心或偏心力。在荷载作用下,这个力能够消除或大大减小跨中关键部位和支座部位的拉应力,从而减小裂缝开展,提高截面的抗弯、抗剪以及抗扭能力。这样,构件能表现出弹性性质,并且当全部荷载作用于结构时,混凝土构件的全部断面的抗压能力都能够被充分有效地发挥出来。

由于钢筋混凝土应变的不兼容性而导致混凝土早期裂缝的发展很可能是发展像预应力混凝土这样新型材料的出发点。

预应力混凝土不是一个新事物,可追溯到 1872 年,当时来自加州的一个工程师

P. H. 杰克森申请了一项预应力系统的专利,他用拉杆把单个的块体建造成了梁或拱。由于在克服预应力损失方面高强度钢筋没有效果,在很长一段时间预应力研究进展很小,亚历山大的 R.E. 迪尔和尼布拉斯卡揭示了混凝土的收缩和徐变(材料横向流变)对预应力损失的影响。在 20 世纪 20 年代早期,美国明尼阿波利斯市的 W.H. 休伊特发展了环向预应力原理。

尤金·弗雷西内于 1926—1928 年间提出了高强度和高延性钢的使用,能克服预应力损失。在 1940 年,他提出了现在众所周知并被普遍认可的弗雷西内预应力法。

预应力混凝土是钢筋混凝土的一个改进形式,钢筋被弯成一定的形状并给它一定的拉力,然后用两种张拉法之一进行预压混凝土。第一种是留下对应钢筋形状混凝土梁的孔道。当钢筋穿过孔道时,梁用水泥浆填满孔道,薄薄的砂浆就与梁黏合在一起。另一种(更常见)方法,是把预应力钢筋放在与成品结构的形状对应的模板的较低部分,并把混凝土浇灌在其周围。这两种方法被称为"先张法"和"后张法"。预应力混凝土因为节省钢材和混凝土,所以是经济、理想的建筑材料。

从前面的讨论中可以清楚地看到,为了消除或大大减少荷载在预应力单元上引起的纯拉应力,在它们承受整个的恒载和活载前,就预先给它们施加一个永久的预压应力。在一般的钢筋混凝土结构中,通常认为混凝土的抗拉强度是可以忽略或不计的,这是因为弯矩产生的拉应力是由加筋处理后的黏合层来抵抗的。因此,钢筋混凝土结构在工作荷载下达到极限状态后产生的裂纹和挠曲变形不可恢复。

和预应力钢筋的作用相反,普通钢筋混凝土构件中的钢筋不对构件施加任何的力。在预应力构件中,钢筋要通过预应力作用给构件主动施加预载,使构件对裂缝和变形有相对较高的恢复控制能力。一旦预应力构件受力使混凝土超过了其抗弯强度,则构件开始表现出钢筋混凝土构件的性质。

在预应力混凝土中,我们应该考虑两种类型的黏结应力。第一种类型可以被认为是"传递黏结应力",而且具有将钢筋的预张力传递给混凝土的功能。第二种类型的黏结力被称为"弯曲黏结应力",当后张构件受到外部载荷时,这种应力存在于预张力和黏结力以及后张构件当中。

当外部荷载变化时,黏结应力也发生在钢筋和混凝土的张拉和黏结当中。当然,在后张构件中,没有传递黏结应力,因为预应力构件端部锚固装置相对较低,故开裂荷载较小,在超出开裂荷载以后,黏结应力有一个突然急剧的增加。因为对超出开裂荷载的混凝土的塑性引起的结果尚不明确,所以在这种情况下,对弯曲黏结应力不能准确计算。此外,设计必须依赖试验的指导。

预应力混凝土使用的钢筋和混凝土较少。由于在受拉区域使用混凝土,和钢筋混凝土相比,可能节省 15%～30% 的混凝土。而钢筋节省会更多,约为 60%～80%,主要是因为高抗拉钢筋的允许应力也高。和钢筋混凝土相比,预应力混凝土尽管能够节省相当数量的材料,但是经济成本并没有显著降低,因为高强混凝土、高强钢筋、锚固以及满足预应力构件生产的硬件要求都会导致附加的费用。尽管有这些附加的费用,通常情况下,如果生产的预制构件在数量上足够多的话,预应力构件和钢筋混凝土构件相比,至少最初的直接成本的差异不是太大。但因为预应力构件不需要太多的维护(因为混凝土质量好,它的使用寿命长),而且由于上部结构的自重较小,基础重量也相应轻得多,所以从长期来看,间接费用的节约还是巨大的。

预应力混凝土的经济性也可以通过大跨度结构得以证明。根据迪恩原理，在美国10～30m长的标准预制桥箱梁和预制预应力桩已经证明比用钢结构和钢筋混凝土结构更经济。据阿贝勒理论，预制预应力混凝土地板、屋顶、跨度达30m的桥梁是经济的；而现浇则对跨度达到100m的结构才更经济。在大跨度结构中，与钢筋混凝土以及钢结构施工相比，通常预应力混凝土更经济。

与其他的施工方式（比如说钢筋混凝土结构和钢结构）相比，预应力混凝土提供了更大的技术优势。以完全预应力构件为例，在工作荷载下不受拉应力，和在工作荷载下带裂缝的钢筋混凝土截面相比，其横截面积可以更有效地利用。在一定的范围内，预应力结构构件中预应力偏心距的增加可抵消永久荷载，从而有效节约使用材料。

预应力混凝土受弯构件在工作荷载作用下比同样高度的钢筋混凝土构件具有更大的刚度。然而，一旦发生开裂，预应力构件和钢筋混凝土构件的弯曲行为相似。因为预压应力的影响，预应力混凝土构件提高了抗剪能力，减少了主拉应力。使用弯曲钢筋，特别是在大跨度结构中，有助于减少在支座截面剪力的发展。

使用高强混凝土和钢筋的预应力构件比钢筋混凝土构件更轻质、更细长。预应力混凝土具有两个特点，即高强度和无裂缝，这归因于在外部侵蚀环境下结构耐久性的提高。混凝土的预应力提高了材料在冲击荷载下对能量吸收的能力，像钢筋混凝土一样，预应力混凝土也具有很高的抵抗重复荷载的能力。

预应力混凝土可用于建造特殊形态的建筑物，像一些没有支柱支撑的大空间的现代体育场馆。这种相对较新的构造方法的使用正在不断发展。

今天，预应力混凝土被用于建筑物、地下结构、电视塔、浮动储藏器、海上结构、电站、核反应堆容器以及包括拱形桥和斜拉桥在内的各种桥梁系统中，这些说明了预应力概念的多方面适应性以及对它的广泛应用。所有这些结构的发展和建造的成功都是由于材料技术进步所获得的无法计量的收获，特别是预应力钢筋和在估计预应力长期和短期损失方面累积的知识。

Lesson 19

What Happens to Structure When the Ground Moves

    This paper focuses on the risk posed to buildings by earthquakes and the steps that can be taken through building regulation and voluntary design education to reduce this risk. First and foremost is the risk to human life in houses, at offices, in schools, in shops and mails, at places of recreation where thousands of people may gather to watch a sporting event or concert, and elsewhere. Beyond the risk to life is the economic and social disruption caused by an earthquake, even moderate earthquakes can result in the loss of many homes, jobs, investments and community resources.

    While earthquakes cause damage and disruption to utilities such as water and power services, these problems are relatively short-lived because utility companies encounter disruption on a normal basis and are equipped to deal with them. Earthquakes may cause severe damage to transportation systems such as railroads and freeways, and collapsing bridges and overpasses may cause injury and death — like that which in the 1989 Loma Prieta earthquake and the 1985 Northridge earthquake in California. These are special problems, however, and need to be dealt with primarily by state transportation agencies. In essence, improving the seismic resistance of buildings is seen as the key to reducing the earthquake threat to the public at large and to the community.

    Issues of health and safety in buildings are typically regulated by building codes written to ensure that some minimal standards of design and construction are adhered to for potentially dangerous aspects of buildings. These codes generally establish such things as maximum loads so that floors of a building will not collapse because they are overloaded with people and equipment and the minimum height of balcony railing so people will not fall over it. These regulations ensure a common minimum standard of safety and mean that building designers work to meet common criteria and do not have to try and solve all the problems of building design on their own every time a new building is planned.

    In regions of the United States such as California and Alaska where earthquakes are frequent, seismic codes have been developed and enforced by local communities many decades, and most existing buildings have been designed with earthquakes in mind. However, since the "science" of earthquake-resistant building design is a relatively new field (the first seismic codes were enforced in California only in 1927), buildings designed to earlier codes are not now necessarily assumed to be safe, and work continues in these regions to, in some instances, strengthen and improve buildings designed to meet the provisions of the earlier codes and to improve the codes.

In regions of the country where the seismic threat has not been accompanied by the continual occurrence of earthquakes, the story is different. ① There may be large inventories of buildings at risk that were designed with no consideration of the seismic problem, and new buildings may still be constructed every year that add to this inventory. When the inevitable large or even moderate earthquake occurs, these buildings may suffer devastating losses. For example, earthquake experts cite the terrible damage to the city of Kobe in Japan where over 5000 people lost their lives in the January 1995 earthquake. This region had been clearly earmarked as an earthquake hazard area by the seismologists and earth scientists, but because a severe earthquake had not affected the city for several hundred years, its buildings (although designed to a seismic code) were vulnerable and its population and government emergency response services were largely unprepared. ②

For communities where a significant earthquake has not occurred in the lifetime of its citizens, the experience of an earthquake is hard to imagine and it is difficult to visualize what an earthquake would do to familiar buildings and other structures. This paper is intended to give readers some idea of the sort of damage that earthquakes do to buildings. The photographs generally show the results of California and Alaska earthquakes and, for the most part, show older buildings designed to lower-than-present-day standards or, in the case of un-reinforced masonry buildings, designed prior to the adoption of seismic codes.

Unreinforced masonry buildings have long been identified as performing very poorly in earthquakes. Unreinforced masonry buildings typically have brick or block bearing walls and wood-framed floors and roofs. The floors and roofs tend to pull away from the walls and collapse; the upper portion of walls, particularly parapets, tend to fall and, depending on the quality and age of the mortar, walls tend to disintegrate.

In California, the state requires that all cities develop an inventory of their un-reinforced masonry buildings and devise a plan for their demolition or improvement. In Los Angeles, an ordinance was enforced in 1981 that required all owners of un-reinforced masonry buildings to demolish or strengthen them. By 1995, essentially all 8000 buildings of this type had either been demolished or strengthened. The 1994 Northridge earthquake showed a notable improvement in the performance of these types of buildings compared to earlier earthquakes—no one was killed and injuries were minimal. San Francisco and a number of other California cities now have similar ordinances in effect.

Older reinforced concrete building structures designed before the characteristics of the material were fully understood have suffered severe damage in earthquakes. Unless heavily reinforced with steel, concrete is a brittle material that lends to fail without warning. In foreign countries, earthquakes caused many total collapses but, in California and Alaska total collapses have been few. Irreparable damage, however, has been significant. Frame structures with few structure walls suffer the most damage, and the problem is less acute for structures with many concrete walls. Seismic codes in force since the 1970s require special reinforcing that greatly reduces the possibility of these brittle failures.

Precast reinforced concrete structures often used for industrial and commercial buildings also have suffered badly in earthquakes. In these types of structures, the damage has been due primarily to inadequate connections between the pre-cast members or between the walls and roof.

The building in the picture (Fig. 19 - 1) is Olive View hospital, which was badly damaged in the 1971 earthquake in San Fernando, California, primarily because of a soft story condition—that is, its lower two floors were much more flexible than the upper floors causing failure where the structure changed from flexible columns to stiff walls.

Fig. 19 - 1   Olive View hospital damaged in the 1971 earthquake

In a typical building the structural components (floor and roof structure, bearing walls, columns, beams, and foundations) account for only about 15 to 20 of the construction cost, the nonstructural architectural, mechanical and electrical components make up between 70 and 85 percent of the building's replacement value.

All these nonstructural components are subject to damage, either directly due to shaking or because of distortion due to movement of the structure. Building occupants are particularly vulnerable to nonstructural damage, and people outside have been injured and even killed by falling parapets and glass. Fires and explosions have been caused by damaged mechanical and electrical equipment. Moreover, nonstructural damage is very costly to repair, and can occur when there is little or no structural damage. It has been estimated that, in recent earthquakes, many buildings with no serious structural damage have suffered considerable nonstructural damage, sometimes totaling as much as 50 percent of the building replacement value.

Ⅰ. New Words

1. earthquake *n.* 地震
2. railroad *n.* 铁路，铁路系统

3. freeway *n.* 高速公路

4. overpass *n.* 立交桥

5. overload *vt.* 使超载，使过载

6. frequent *adj.* 频繁的，时常发生的，惯常的

7. devastate *vt.* 毁灭，破坏

8. seismologist *n.* 地震学家

9. collapse *vi.* 倒塌，瓦解

10. parapet *n.* 栏杆，扶手

11. demolition *n.* 拆除，拆卸

12. strengthen *vt.* 加强，加固

13. brittle *adj.* 脆性的，易碎的，脆弱的

14. irreparable *adj.* 无法弥补的，不能挽回的

15. flexible *adj.* 柔性的，柔韧的，易弯曲的

Ⅱ. Phrases and Expressions

1. building regulation　建筑规范，建筑规程
2. seismic resistance　抗震性
3. earthquake hazard area　地震灾害区
4. brittle material　脆性材料
5. standard of design　设计标准
6. precast reinforced concrete structure　预制钢筋混凝土结构
7. flexible column　柔性柱
8. stiff wall　刚性墙
9. nonstructural component　非结构构件

Ⅲ. Notes

① "In regions of . . ."在句子中作状语，但其中包含了 where 引导的定语从句，真正的主语是后面的"the story"。

② 此句较长，要分清句子成分有利于理解文中的意思。其中"its"指代"this region"，译为"该地区的"。即"its buildings"译为"该地区的建筑物"，"its population"译为"该地区的人口"。

Ⅳ. Exercises

Fill in the blanks with the information given in the text.

1. Earthquakes may cause severe damage to _____ systems such as _____ and _____, and collapsing _____ and _____ may cause injury and death—like that which in the 1989 Loma Prieta earthquake and the 1985 Northridge earthquake in California.

2. Issues of health and safety in buildings are typically regulated by _____ written

to ensure that some _____ standards of design and construction are adhered to for dangerous aspects of buildings.

3. Building occupants are particularly vulnerable to _____ damage, and people outside have been injured and even killed by falling _____ and _____ .

Ⅴ. Expanding

Know about the following terms related to the earthquake and seismic design.
1. hypocenter 震源
2. epicenter 震中
3. epicentral distance 震中距
4. epicentral region 震中区
5. depth of seismic focus 震源深度
6. earthquake/seismic intensity 地震烈度
7. basic earthquake intensity 基本地震烈度
8. fortification intensity 设防烈度
9. earthquake wave 地震波
10. building load code 建筑荷载规范
11. seismic collapse resistance 抗地震倒塌
12. concept design for seismic-resistance 抗震概念设计

Ⅵ. Reading Material

Earthquake

Earthquakes are vibratory phenomena associated with shock loading on the earth's crust, while these shock loads can result from a number of causes, one of the primary reasons is the sudden slippage that frequently occurs between adjacent crust plates that make up the earth's surface.

Most earthquakes occur within the upper 15 miles of the earth's surface. But earthquakes can and do occur at all depths to about 450 miles. Their number decreases as the depth increase. At about 460 miles one earthquake occurs only every few years. Near the surface, earthquakes may run as high as 100 in a month, but the yearly average does not very much. In comparison with the total number of earthquakes each year, the number of disastrous earthquakes is very small. Examples of such disastrous earthquakes are the 1999 Izmir, Turkey earthquake; the 1999 Jiji, Taiwan China Earthquake; the 1995 Henshin Japan Earthquake; the 1994 Northridge, California U. S. Earthquake; the 1976 Tangshan, China Earthquake and many others.

China is one of the most seismically active regions in the world. There have been about 300 earthquakes with magnitudes greater than six in the continent of China since 1900 and seven of these have had magnitudes greater than eight. The largest earthquakes in China generally occur in one of five zones: (1) the Himalayan zone; (2) the central Asia zone, extending northeast from Pamir, through Altai in western Mongolia to Baikal; (3) the north-south zone, extending along the eastern margin of the Qinhai-Tibet Plateau; (4) the north China plain zone, which includes the Fenwei zone, the Hebei Plain and the Tanlu zone, along the Pacific Ocean.

The extent of the disaster in an earthquake depends on many factors. If you carefully build a toy house with an Erector set, it will still stand to matter how much you shake the table. But if you build a toy house with a pack of cards, a slight shake of the table will make it fall. An earthquake in Agadir, Morocco, was not strong enough to be recorded on distant instruments, but it completely destroyed the city. Many stronger earthquakes have done comparatively little damage. If a building is well constructed and build on solid ground, it will resist on earthquake. Most deaths in earthquakes have been due to faulty building construction or poor building sites. A third and very serious factor is panic. When people rush into narrow streets, more deaths will result.

The United Nations has played an important part in reducing the damage done by earthquakes. It has sent a team of experts to all countries known to be affected by earthquakes. Working with local geologists and engineers, the experts have studied the nature of ground and the type of most practical building code for the local area. If followed, these suggestions will make disastrous earthquakes almost a thing of the past.

There is one type of earthquake disasters that little can be done about. This is the disaster caused by seismic sea waves, or tsunamis. In certain area, earthquakes take place beneath the sea. These submarine earthquakes sometimes give rise to seismic sea waves. The waves are not noticeable out at sea because of their long wave length. But when they roll into harbors, they pile up into walls of water 6 to 60 feet high. The Japanese call them "tsunamis", meaning "harbor wave", because they reach a sizable height only in harbors.

Tsunamis travel fairly slowly, at speed up to 500 miles an hour. An earthquake warning system is in use to warn all shores likely to be reached by waves. But this only enables people to leave the threatened shores for higher ground. There is no way to stop the oncoming wave.

In spite of the great progress obtained in the field of earthquakes engineering during the past fifty years, recent destructive earthquakes occurred around the world revealed that, the existing knowledge and techniques are still not sufficient to achieve safety against earthquakes at an effective cost. It is believed that among all natural hazards earthquakes are still number one disaster for which in-depth research, particularly for those devastating earthquakes, the basic research on earthquake engineering is still in need to expand our knowledge and strengthen our defenses.

In recognition of the recent rapid advancement of technologies related to earthquake engineering, high-speed development of satellite remote sensing technology has played significant roles in reducing various kinds of natural disasters, it can be used in rapid assessing the seismic damage for effective post quake emergency action and in monitoring crustal movement for better understanding of seismic risk. The Digital Disaster Reduction System would be a specially designed system to study the virtual seismic damages that may happen to real structures during real earthquakes. It is a virtual reality computer system designed to simulate the occurrence and propagation of disaster and whole process of damages caused by natural disasters. The Digital Disaster Reduction System could be applied as a powerful tool not only for seismic disasters study but also for other natural disaster research.

参 考 译 文

第 19 课　地层运动时，结构会发生什么

本文集中讨论地震在建筑物上引起的危险，通过建筑规范与自主性设计培养后能采取措施以降低风险。首要的是，处于住宅、学校、商场及可能有数千人聚集观看体育比赛或听音乐会的休闲场所，以及别的地方的人的生命危险。除了人的生命危险之外，就是地震引起的经济和社会破坏，即使中强地震都会造成许多家庭、工作、投资和社会资源的损失。

尽管地震会引起公共设施如供水和供电设施的破坏和中断，但这些问题是相对短期的，因为一般情况下公共设施公司都会遇到这样的中断，而且配有应对装备。地震可能引起交通系统如铁路和高速公路的严重破坏，而且桥梁和天桥的坍塌会造成人员伤亡，就像1989年洛马普列塔地震和1985年加州北岭地震中的那样。然而，这些是特殊问题主要由州交通部门处理。实质上，提高建筑物的抗震能力被看成是减少地震对公众和社会威胁的关键。

建筑物的健康与安全问题，一般可通过书面的建筑规范控制，以保证一些基本设计与施工标准得到坚持，以避免建筑物的某些潜在危险。这些规范一般建立在最大载荷这样的事情上，使得建筑楼板不会因为人与设备超载而垮塌，保证走廊栏杆的最小高度使人们不会翻过跌落。这些规定保证通常的最低安全标准，这意味着建筑设计者要努力工作来满足通常的标准要求，而不必努力在每次设计新建筑时依靠自己去解决建筑设计中的所有问题。

在美国的地震多发区，如加利福尼亚和阿拉斯加，抗震规范得到了发展，几十年来被当地社会执行，现有的大多数建筑物设计时都考虑了抗震要求。然而，由于抗震建筑设计"科学"是相对较新的领域（第一个抗震规范于1972年在加利福尼亚才得以执行），按照早期规范设计的建筑现在认为是不一定安全的，某些情况下在这些地区还必须继续工作，以加固和改善依照早期规范条款设计的建筑，并且完善规范。

在有些国家存在地震威胁但没有频繁发生地震的地区，情况则不相同。许多处于危险中的建筑物数目繁多，这些建筑物设计时未考虑抗震问题，而且每年新建的建筑物一直都在加长这一危险建筑物清单。当不可避免的大地震或中度地震出现时，这些建筑物可能遭受灾难性损失。如地震专家引述日本神户市的骇人破坏，1995年1月的那场地震导致了5000多人失去生命。该地区过去一直被地震专家和地球科学家认定为地震危险区，但由于几百年来严重的地震未影响到神户市，当地的建筑物（虽然设计时考虑了抗震规范）是

脆弱的，当地人们和政府的紧急响应服务在很大程度上处于无准备状态。

在其居民一生都未出现大地震的社区，地震经历是难以想象的，而且也难以设想地震对熟悉的建筑物和其他结构物造成的后果。本文试图给读者有关地震破坏建筑物的方式的一些概念，图片大致展示了加利福尼亚和阿拉斯加地震的后果，而且在极大程度上显示了按照低于现行抗震规范标准设计的旧建筑物，或抗震规范执行前未进行抗震加强的砖混建筑物的破坏结果。

长期以来，人们认为未进行抗震加强的砌体建筑的抗震性能是很差的。典型的未进行抗震加强的砌体建筑有砖或砌块承重墙与木框架的楼板和屋顶。楼板和屋顶趋向于与墙体脱离和倒塌，墙体上部特别是女儿墙趋于垮塌，而且根据砂浆的质量和寿命，墙体也趋于破坏。

在加利福尼亚，州政府要求所有城市都要提出未抗震加强的砌体建筑的清单，并提出对这些建筑物进行拆除或加固计划。在洛杉矶，1981年生效了一项法令，要求所有抗震加固建筑物的所有者于1995年前要么拆除，要么加固建筑物，实质上是要对所有8000幢这类的建筑物或者拆除，或者加固。1994年的北岭地震表明，与早期的地震相比，这类建筑物的性能得到了明显改进——没有人员死亡，受伤的人很少。旧金山和其他的很多加利福尼亚城市现在也生效了类似的法令。

人们对材料特性完全了解之前设计的旧钢筋混凝土建筑结构，在地震中遭受了严重破坏。如果没有用钢筋高度加强，混凝土是脆性材料，会导致毫无预兆的破坏。在国外，地震造成了许多建筑彻底垮塌，但是在加利福尼亚和阿拉斯加，彻底的垮塌很少。然而，那些不可挽回的损失是巨大的。几乎没有构造墙体的框架结构遭到了最严重的破坏，而对有很多混凝土墙的结构物，问题则没那么突出。自从20世纪70年代生效的抗震规范要求专门的加固后，这些脆性破坏的可能性大大降低了。

通常用于工业和商业厂房的预制钢筋混凝土建筑物也在地震中遭受严重破坏，在这些类型的建筑物中，破坏主要是由于预制构件之间或墙体与屋顶之间的连接不足。

图19-1中的建筑物是奥利弗医院，在1971年加利福尼亚旧金山地震中遭受了严重破坏，原因主要在"薄弱层"（即其下部两层的柔性大于引起破坏的上部楼层）条件下引起破坏，这种破坏发生在结构从柔性柱变成刚性墙的地方。

图19-1 1971年地震中损坏的奥利弗医院

在典型的建筑物中，结构构件（楼板和屋面结构、承重墙、柱、梁、基础）仅占到建筑成本的15%～20%，非结构的建筑、机械和电器构配件占据建筑物重置价值的70%～85%。

所有非结构构件都受到了损坏，或者直接由于摇晃，或者由于结构建筑运动引起的扭曲。房屋的居住者特别易受到非结构性构件破坏的伤害，垮落的女儿墙和玻璃则易造成屋外人的伤害，甚至死亡。受损的机械和电器部件会引起火灾和爆炸，此外，修复损坏的非结构构件是很昂贵的，而且可能发生在没有或有少量的结构损坏的时候。据估计，在最近的地震中，没有严重结构损坏的很多建筑物都遭受了相当大的非结构性损坏，有时总损失高达建筑物重置价值的50%。

Lesson 20
Underground Space Utilization

The rapid growth of world civilization will have a significant impact on the way humans live in the future. As the global population increases and more countries demand a higher standard of living, the world must provide more food and greater energy and mineral resources to sustain this growth. The difficulty of doing this is compounded by three broad trends: the conversion of agricultural land to development uses; the increasing urbanization of the world's population; and growing concern for the maintenance and improvement of the environment, especially regarding global warming and the impact of population growth. Underground space utilization, as this lesson describes, offers opportunities for helping address these trends.

By moving certain facilities and functions underground, surface land in urban areas can be used more effectively, thus freeing space for agricultural and recreational purposes. Similarly, the use of terraced earth sheltered housing[①] on steeply sloping hillsides can help preserve precious arable flat land in mountainous regions. Using underground space also enables humans to live more comfortably in densely populated areas while improving the quality of live.

On an urban or local level, the use of underground facilities is rising to accommodate the complex demands of today's society while improving the environment. For example, both urban and rural areas are requiring improved transportation, utility, and recreational services. The state of traffic congestion in many urban areas of the world is at a critical level for the support of basic human living, and it is difficult if not impossible to add new infrastructure at ground level without causing an unacceptable deterioration of the surface environment or an unacceptable relocation of existing land uses and neighborhoods.

On a national level in countries around the world, global trends are causing the creation and extension of mining developments and oil or gas recovery at greater depths and in more inaccessible or sensitive locations. These trends have also led to the development of improved designs for energy generation and storage systems as well as national facilities for dealing with hazardous waste (including chemical, biological, and radioactive waste), and improved high-speed national transportation systems. All these developments involve use of the underground.

1. Land use pressures

Placing facilities underground is a promising method for helping ease land use pressures caused by the growth and urbanization of the world's population. Although the average population density in the world is not large, the distribution of population is very uneven. A map of population density indicates that large areas of the world are essentially uninhabited. These areas are for the most part deserts, mountainous regions, or regions of severe cold that do not easily support human habitation.

If one examines China, for example, the average population density is approximately 100 persons per square kilometer, but the vast majority of the one billion-plus population lives on less than 20 percent of the land area. This is the fertile land that can support food production. However, due to population growth, urbanization, and economic growth, this same land must now support extensive transportation systems, industrial and commercial developments, and increasing demands for housing. As the population and economy grow, the land available for agriculture shrinks, and the problems of transporting food and raw materials to an urban population increase. By the year 2000 it is estimated that 70 percent of the world's population will inhabit urban areas.

The same trends are evident in Japan, where approximately 80 percent of the land area is mountainous, 90 percent of the population lives on the coastal plains, and economic development is concentrated in relatively few economic centers. The flat-lying land is generally the most fertile and is historically the region of settlement. Other factors adding to population density include the traditional building style, which is low-rise, and Japanese laws that contain strong provisions for maintenance of access to sunlight. Also, to retain domestic food production capability, the Japanese government has protected agricultural land from development. The combination of these historical and political factors together with a strong migration of businesses and individuals to the economic centers has created enormous land use pressures. The result is an astronomically high cost of land in city centers (as high as US $ 500000 per square meter) and difficulty in providing housing, transportation, and utility services for the population. Typical business employees cannot afford to live near the city center where they work and may have to commute one to two hours each way from an affordable area. To service the expanding metropolitan area, public agencies must upgrade roads and build new transit lines and utilities. Land costs for such work are so high that in central Tokyo, the cost of land may represent over 95 percent of the total cost of a project.

The problem of land use pressures and related economic effects of high land prices are of great interest in the study of the potential uses of underground space. When surface space is fully utilized underground space becomes one of the few development zones available. It offers the possibility of the adding needed facilities without further degrading the surface environment. Without high land prices, however, the generally higher cost of constructing facilities underground is a significant deterrent to their use. When underground facilities are not economically competitive, they must be justified on aesthetic, environmental or social grounds which many developing nation cannot afford at present and which developed nations are reluctant to undertake except in areas of special significance.

2. Planning of underground space

Effective planning for underground utilization should be an essential precursor to the development of major underground facilities. This planning must consider long-term needs while providing a framework for reforming urban areas into desirable and effective environments in which to live and work. If underground development is to provide the most valuable long-term benefit possible then effective planning of this resource must be conducted. Unfortunately, it is already too late for the near-surface zones beneath public rights-of-way in older cities around the world. The tangled web of utilities commonly found is due to a lack of coordination and the historical evolution in utility provision and transit system development.

The underground has several characteristics that make good planning especially problematical:

(1) Once underground excavations are made the ground is permanently altered. Underground structures are not as easily dismantled as surface buildings.

(2) An underground excavation may effectively reserve a large zone of the stability of the excavation.

(3) The underground geologic structure greatly affects the type size and costs of facilities that can be constructed but the knowledge of a region's subsurface can only be inferred from a limited number of site investigation borings and previous records.

(4) Large underground projects may require massive investments with relatively high risks of construction problem delay and cost overruns.

(5) Traditional planning techniques have focused on two-dimensional representations of regions and urban areas. This is generally adequate for surface and aboveground construction but it is not adequate for the complex three-dimensional geology and built structures often found underground. Representation of this three-dimensional information in a form that can readily be interpreted for planning and evaluation is very difficult.

In Tokyo, for example, the first subway line Ginza Line was installed as a shallow line 10 meters deep immediately beneath the existing layer of surface utilities. As more subway lines have been added uncluttered zones can only be found at the deeper underground levels. The new Keiyo JR line in Tokyo is 40-meter deep. A new underground super highway from Marunouchi to Shinjuku has been proposed at a 50-meter depth. For comparison the deepest installations in London are at approximately a 70-meter depth although the main complex of works and sewers is at less than 25 meters. Compounding these issues of increasing demand is the fact newer transportation services (such as the Japanese Shinkansen bullet trains or the French TGV) often require larger cross-section tunnels straighter alignments and flatter grades. If space is not reserved for this type of use very inefficient layouts of the beneath urban areas can occur.

3. Environmental benefits

Another major trigger for underground space usage is the growing international concern over the environment which has led to attempts to rethink the future of urban and industrial development. The major concerns in balancing economic development versus environmental degradation and world natural resource limitations revolve around several key issues. These are:

(1) The increasing consumption of energy compared to the limited reserves of fossil fuels available to meet future demand.

(2) The effect on the global climate of burning fossil fuels.

(3) The pollution of the environment from the by-products of industrial development.

(4) The safe disposal of hazardous wastes generated by industrial and military activities.

Preserving the environment and extending the life of the world's resources while promoting economic growth and maintaining individual life styles will be complex if not impossible. However a high standard of living and high gross domestic product do not have to be proportionately dependent on resource consumption and environmental degradation.

Underground space utilization can help solve the environmental/resource dilemma in several ways. Underground facilities are typically energy conserving in their own right. More importantly by using underground space, higher urban densities can be supported with less impact on the local environment. In addition to the obvious benefit of preserving green space and agricultural land there is strong evidence that higher urban density can lower fuel resource consumption.

4. The future of underground space development

Although existing underground facilities throughout the world provide some models for future development they are all limited in scale, in use, or in their lack of a comprehensive vision for the total city environment. As a complement to more detailed planning and research studies it is useful to examine the visions of extensive underground complexes even entire cities that have been proposed by futuristic planners and designers.

Geotech'90, a conference and exhibition held in Tokyo in April 1990, was a major forum for the underground industry in Japan. More than a dozen underground concepts were displayed ranging from the typical transit and utility uses to underground corridors that are envisioned as places for a communication network protected during disasters. Such corridors could also effectively transport both waste and energy between substations in the city and central generation and disposal sites outside the city. This approach not only relieves congestion but also can provide more efficient energy generation and recycling of waste materials. These concepts are all intended to permit a major upgrade of the city infrastructure that will eventually enable the surface to be rebuilt with more open space and a more efficient attractive overall environment.

When completely new cities are envisioned for the future the underground often is a major component as illustrated by the work of the architect Paolo Soleri[②] over the last 30 years. In science fiction future cities often are depicted as self-contained climate-controlled units frequently located underground for protection from the elements and possibly from a hazardous or polluted environment. In this case underground cities on earth differ little from bases created on the moon or other isolated environments.

Ⅰ. New Words

1. conversion *n.* 转变，转化
2. terrace *vt.* 使成阶地；*n.* 平台，阶地，阳台
3. congestion *n.* 拥挤，拥塞
4. infrastructure *n.* 基础设施，下部结构
5. deterioration *n.* 退化，恶化
6. uninhabited *adj.* 无人居住的，杳无人迹的
7. habitation *n.* 居住，住处，住宅
8. metropolitan *adj.* 主要城市的，大都市的
9. deterrent *n.* 制止物，威慑，阻碍
10. precursor *n.* 先驱，前导，前辈
11. tangle *vt.* 使纠缠，使困惑，使混乱
12. excavation *n.* 挖掘，开凿，洞穴
13. dismantle *vt.* 拆开，拆除，拆卸
14. alignment *n.* 排列，直线排列，校准
15. trigger *n.* 引发物，引爆器，扳机
16. versus *prep.* 对，对抗，与……比较
17. degradation *n.* 退化，降级，降解
18. forum *n.* 论坛，讨论会，会场
19. corridor *n.* 走廊，通道，过道

Ⅱ. Phrases and Expressions

1. global warming　全球变暖
2. impact of population growth　人口增长的影响
3. average population density　平均人口密度
4. coastal plain　海岸平原

Ⅲ. Notes

① "earth sheltered housing" 译为 "掩体住宅建筑"，此处指在陡峭的山坡上利用阶地建造的掩土住宅，它具有可以有效利用地热能、节约能源、高效利用城市土地、维修成本低等优点。

② 保罗·索拉尼（Paolo Soleri）是意大利裔美国籍城市规划家、规划理论家和建筑

师。他的"建筑生态"观集中体现在位于亚利桑那州的试验城镇"阿科桑蒂（Arcosanti）"的规划和设计上，给当代城市规划的生态化、人际化、低水平维养、高密度居住提供了一个很好的范例和课题。

Ⅳ. Exercises

Answer the following question according to the text.

1. Why is the underground space utilized so widely at present?
2. Please list the characteristics of underground that make good planning especially problematical.
3. What are the key issues in balancing economic development versus environmental degradation and world natural resource limitations?

Ⅴ. Expanding

Know about the following terms in relation to the underground structures.

1. utility tunnel　综合管廊
2. trunk utility tunnel　干线综合管廊
3. branch utility tunnel　支线综合管廊
4. cable trench　缆线管廊
5. airdefence basement　防空地下室
6. immersed structure　沉管结构
7. open caisson structure　沉井结构
8. diaphragm wall　地下连续墙
9. top-down construction　逆作法施工

Ⅵ. Reading Material

Subway Engineering

Subway engineering is a branch of transportation relating to planning, general layout, detailed design, construction, and operation of subways. These underground systems are a major element in mass transportation. As metropolitan areas increase in size and population and vehicular traffic becomes more congested, subways are being given increased consideration in planning urban mass transit systems.

The first subway opened in London in 1863. Steam locomotives fueled by coke and coal were used to pull the trolley cars. In 1896 the first subway on the European continent was placed in service. In Boston, Massachusetts, subway lines were instituted in 1895 and 1897. Since then, subways have been constructed in several United States cities.

The most modern subway trains are designed for high-speed travel with maximum safety in comfortable air-conditioned cars. Train movements are controlled by automatic equipment. Since construction of a subway involves putting a railway of special design underground, most of the engineering specialists relating to railways are required in building a subway.

However, in subway work there is an emphasis on extensive and costly tunnel construction, with tunnels often being at sufficient depth to pass under bodies of water; underground passenger terminals; escalators for transporting passengers to and from street level; noise control; and safe and reliable signaling facilities.

1. Planning

Engineering planning as related to subways includes comprehensive studies to determine whether a subway system is economically feasible. This work involves extensive analysis to evaluate forecasts of construction costs, passenger volumes, passenger fares and revenue, operating costs, depreciation and maintenance of equipment, and passenger safety.

2. Construction

Often difficult to install, subways are constructed by either the open-cut method, the tunneling method, or both.

With the open-cut method, a deep open trench is excavated. This frequently requires shoring and bracing at both sides of the trench, in addition to heavy cross framing at different levels. Open-trench procedures limit the depth at which work can be performed.

Tunneling permits subways to be installed at great depths. However, the operation can be costly due to the type of soil or foundation material encountered, unstable conditions, and extensive flows of water.

The work of building subways is further complicated by existing underground water supply mains, storm drainage facilities, sanitary sewers, and conduits for electric services and telephone lines. Frequent vehicular traffic along the street under which the subway must be located may add further problems. Often, it is difficult to traffic on a busy roadway and to close the route to traffic for a long period during subway construction. In such cases a modified open-trench procedure may be followed. Heavy steel vertical pilings are driven along the sides of the area where work must be done. Strong horizontal timbers or sheeting are inserted behind the pilings, and heavy transverse steel beams are installed between the pilings at frequent intervals. Strong timbers or concrete members are then placed longitudinally between the steel beams to serve as a temporary roadway to permit resumption of street traffic. Because of the large space needs for the subway tubes, trucks and other equipment can easily work below the temporary roadway to excavate and haul away material. This procedure has been used for projects in the United States, in building the subway in Toronto, Ontario, and for a subway in West Berlin.

3. Underpinning

Another factor that makes subway construction difficult is nearby buildings and other structures, especially if these are of sufficient weight to cause heavy loads on the underlying foundation material and large pressures in the areas of the proposed subway. Frequently, heavy underpinning must be installed to support adjacent buildings during subway construction. This underpinning may consist of steel pilings and heavy horizontal members on which buildings can be supported. Large concrete columns reinforced with steel also are used as supports; columns that are as large as 4—5 ft. (1.2—1.5 m) in diameter have been used. During the installation of any underpinning and the following subway construction activity, it is generally necessary to reduce vibration and settlement of foundation materials to a minimum.

The design of the Bay Area Rapid Transit (BART) system included a difficult but satisfactory method to provide subway service under the harbor between San Francisco and Oakland. Dry dork sections about 300 ft. (90 m) long made of large-diameter concrete tubes were constructed. These sections were built with temporary closures at the ends so that they could be floated out into the bay and lowered with large rigs into a previously prepared trench in the bay bottom. Each section was joined to the last one placed. Once the sections were in position, the temporary closures at the ends were removed. The procedures in building the BART system were so successful that they were adopted for constructing a 1 mi (1.6 km) section of subway under the Hong Kong harbor.

Determination of the best location for a subway requires much careful study and consideration of various general plans, including underpinning. In some metropolitan areas the subway is built underground in the central business area and aboveground at other locations. The subway in the southern part of Chicago was placed along an interstate highway at ground level in the median strip, the space dividing opposing traffic on the highway. On another Chicago line leading westward from the central business district, the subway was constructed in a tunnel below the median strip of the Congress Street Expressway.

4. New systems

Because of "downtown" congestion and the difficulty of providing adequate parking, several United States cities have constructed rapid transit facilities. One major subway system is the 98 mi (156 km) network serving Washington D.C. Extensive auto parking facilities near the several terminals are an integral part of the project. In Atlanta the 54 mi (86 km) rail transportation system includes many miles of bus routes.

New subway systems or major additions to older facilities are planned or have been placed in service in several other American cities, including Baltimore, Houston, Los Angeles, Philadelphia, and Pittsburgh.

Numerous foreign countries have completed or have under construction subway systems. Many foreign countries are served by urban subway networks. These include Argentina, Australia, Austria, Belgium, Brazil, Canada, Chile, China, Czechoslovakia, Denmark, England, Finland, France, Greece, Hungary, India, Ireland, Italy, Israel, Japan, Mexico, the Netherlands, Republic of Korea, Norway, Portugal, Romania, Scotland, the Former Soviet Union, Spain, Sweden, Switzerland, and Germany.

Some of the foreign systems under construction are large comprehensive projects of considerable mileage and great cost. For example, the network under construction for the island of Singapore, which has an area of only 220 mi^2 (570 km^2) and a population of 2.5 million, will have a length of 42 mi (67 km) with 42 passenger stations and will cost US＄2.3 billion. Scheduled to go into service in 1988, nearly a third of the system is being built underground by tunneling methods.

As of 1985 a total of nearly 2400 mi (3800 km) of subway was in operation throughout the world. By the year 2000, the lines in service are expected to total over 5400 mi (8600 km).

参 考 译 文

第20课　地下空间的利用

　　全球城市化进程的加快将会对人类将来的生存方式产生重大影响。随着全球人口的增长以及更多国家要求提高生活水平，世界必须提供更多的食物、能源以及矿物资源来维持这种增长趋势。解决这一难题的办法由三大趋势复合而成：为了更深入利用而进行的农业用地的保护；世界人口的日益增长；对保护和改善环境日益增长的关注，特别是关于全球气候变暖以及人口增长带来的影响。地下空间的利用，作为本课要描述的内容，将提供针对这些趋势的解决办法。

　　通过将特殊器材设备置于地下，城市地表可被更有效地利用，这样就可以为农业和休闲释放出空间。同样地，在陡峭的山坡上使用阶地掩土住宅会有助于在多山地区保护宝贵的可耕平地。利用地下空间也可以使人们在改善生活质量的同时，提高人们在人口高密集区居住的舒适度。

　　以城市或当地水准，地下设施的利用在改善环境的同时正日益满足当今社会复杂的需求。例如无论是城市还是农村都需要改善运输、公用事业和娱乐服务。世界上许多城市的交通堵塞现象已经处在满足人类基本生存需求的临界点上，并且在不破坏地表环境的基础上不增加新设施或不重新规划现有土地及周边地带上的建筑的基础上，想要解决这一难题是十分困难的。

　　以世界各国的国家水平，全球化的趋势导致对采矿、石油或天然气的开采等已到达更深的地层以下，并进一步延伸，触及更难以让人接受或是更敏感的区域。这些趋势同样导致针对能源的产生与存储系统和用于处理危险废料（包括化学、生物以及放射性废弃物）的国家设施设计的改善发展，同样也改善了国家高速运输系统。所有的这些发展均涉及地下空间的利用。

1. 用地压力

　　将设施置于地下是帮助缓解由于世界人口的增长和城市化所带来的地上用地压力问题的一种有前途的办法。虽然世界平均人口密度并不大，但人口分布却很不均匀。世界人口密度图显示世界上大部分地方根本不适合居住。这些地方大部分是沙漠、山区，或是极度严寒的地带等人类不易居住的地区。

Lesson 20 Underground Space Utilization

以中国为例，平均人口密度大概是每平方公里100人，但是10亿多的绝大部分人口居住在不到20%的国土上。这是那些可以提供粮食产品的肥沃土地。然而，由于人口增长、城市化和经济增长，这些土地现在同时还要支撑广大的运输系统、工商业的发展，以及日益增长的住房需求。随着人口和经济的增长，农业用地逐渐减少，向城市人口运送食物和原材料的问题日益增长。据估计，到2000年，世界人口的70%将居住在城市。

同样的趋势在日本也很明显，大约80%的陆地是山区，90%的人口居住在海边平原地区，经济的发展主要集中在相对较少的经济中心。平坦的陆地地区通常是最肥沃的，从历史上看也是人类的定居地。增加人口密度的其他因素还包括：传统的低层建筑模式，而且日本法律规定必须建造有足够的光照的坚固的维护设施。同样，为了保持国内的粮食生产能力，日本政府从发展的角度已经开始保护农业用地。这些历史、政策因素的结合，以及商业和个人均向经济中心大量的移民等造成了巨大的土地使用压力。结果就是市中心土地惊人的高价（高达每平方米50万美元），以及很难为人们提供住房、交通、公共设施服务。普通公司雇员无法承担居住在他们工作的市中心附近，而不得不从他们能负担得起的住处搭乘公共车辆单程花1~2个小时到公司。为了服务于日益扩大的大城市地区，政府当局必须升级道路、兴建新的交通线路和公共设施。这些工程的用地费用如此昂贵以至于在东京市中心用于购买土地的费用可能会占到工程总费用的95%以上。

土地使用压力和与之相关的高土地使用价格的经济影响使得人们对地下空间的潜在利用的研究产生了极大的兴趣。当地表土地已被利用殆尽，地下空间将变成可开发的区域之一。这为不需要深度破坏地表环境而又能增加需要的设施提供了可能性。然而，在通常情况下，虽然不需高额的土地价格，但是建造地下设施的高额花费却是地下空间利用的一大障碍。当地下设施不具有经济竞争力时，必须在美学、环境或社会因素等方面对它们（地下设施）进行论证，除非具有特殊的重大意义，否则这些方面很多发展中国家目前还承担不起，而发达国家又不愿意承担。

2. 地下空间规划

对地下空间利用的有效规划是较多地下设施发展的必要前奏。在为把城区改造成称心的和有效的生活与工作环境提供规划框架时，这种规划必须考虑（城市）长期发展的需要。如果地下空间开发可以提供最具价值的可能的长期效益，那么对这些资源的有效计划就应得以实施。令人遗憾的是，在世界上较老的城市中，在公共路权下的近地表区域开发已经太晚了。通常，紊乱的公共设施网络归咎于在公共设施供给方面和交通系统发展方面缺乏协调以及历史性沿革等原因。

地下空间具有以下几个特征，需要我们做好规划，尤其是有问题的地方。

（1）一旦开始地下开挖，土地将被永久改变。地下建筑不像地面建筑那样容易拆除。

（2）一处地下空间的开挖实际上可能需要保留一大片开挖稳定区域。

（3）土地的地质构造极大地影响了拟建地下设施的类型、规格以及费用。然而，关于一个区域地表下土壤（或岩石）的知识（或具体情况）仅能根据有限数量的现场钻探资料和以前的记录来推测。

（4）大型地下工程项目可能需要对施工难题、工期拖延、预算超支等这些相对高风险的问题进行大量的调查。

（5）传统的规划技术主要侧重于对于各地区以及城市区域的二维描述。这基本上仅适合地表及上部结构，但并不适合建造在处于复杂三维地理环境中的地下结构。用一种容易对规划和评估进行解释的模式来对这种三维信息进行描述是非常困难的。

例如，在东京，第一条地铁线（Ginza 线）是在已建地面公共设施的地表层下作为一个浅层线路（10m 深）建造的。随着更多地铁线的增设，只有在更深的土层中才会发现比较规整的区域。在东京，新的地铁线 Keiyo JR 线深达 40m。一条新的从 Marunouchi 到 Shinjuku 的地下高速公路已被设计到 50m 深。作为对比，在伦敦，尽管其主要的工程和污水管道系统的复杂部分都在不超过 25m 处，但最深的设施已达约 70m 深。综合日益增长的各种需要后，得到这样一个事实，就是新型交通运输服务（如日本的新干线子弹头列车或是法国的高速列车）通常需要的隧道应具有较大的横断面、较笔直的平面布置和较平坦的坡度。如果地下空间不是用作此种类型的用途，那么城市地下将会产生非常无效率的布局。

3. 环境利益

地下空间利用的另一个主要激发因素是国际上对环境问题越来越多的关注，它使我们尝试去重新考虑城市和工业发展的未来。对平衡经济发展对抗环境退化和世界自然资源有限的主要关注点是围绕以下几个关键问题展开的。它们是：

（1）越来越多的能源消耗，与能满足将来所需的矿物燃料的有限储备量之间的矛盾；
（2）燃烧矿物燃料对全球气候带来的影响；
（3）工业生产的副产品对环境的污染；
（4）对于工业及军事活动产生的有害废弃物的安全处置。

在促进经济增长、保持个人生活模式的同时，保护环境、延长地球上资源的寿命，如果可能的话，也是很复杂的。然而，高的生活标准和高国内生产总值（GDP）不必要相应地依赖资源的消耗和环境的退化。

地下空间的利用能从几种途径帮助解决环境/资源的困境。典型的是，地下设施凭其本身的质量，可做到能源节省。更重要的是，通过使用地下空间，在对当地环境产生较小影响的情况下，就能支撑较高的城市密度。除了能保护绿色空间和农业用地这明显的益处外，有力的证据表明，较高的城市密度可以减少燃料资源的消耗。

4. 地下空间发展的未来

虽然全世界已有的地下设施提供了未来发展的一些模式，但是它们在规模上，在使用上，或是在对于整个城市环境缺乏全面的视角方面等，都是受限的。作为对更详细的规划和调查研究的补充，调查庞大的地下综合体甚至整座城市的远景是有用的，这已经被未来的规划者和设计者提出来了。

Geotech'90——1990 年 4 月在日本东京举行的一个研讨博览会，是日本地下工业的一个重要论坛。有许多地下的观念被展示出来——范围从典型的运输和公用设施的使用到灾难期间被想象作为保护通信网络场地的地下走廊。这类走廊还能够有效地在城内的变电站和城外的主要生产和处理场所之间输送废弃物和能源。这种方法不仅缓解了拥挤而且提供了更有效的能源产生和废物的再循环利用。这些观念都将容许城市基础设施的一次重要升级，最后使地表能被重建得具有更开阔的空间和更有效、更吸引人的全面的环境。

Lesson 20　Underground Space Utilization

当展望未来全新的城市的时候，通常地下就成为一个重要的组成部分，就像过去 30 多年前由建筑师保罗·索列里的作品展现的一样。在科幻小说中，常把未来的城市描绘成设备齐全、气候可控的单元，这些单元经常处于地下，以遮蔽风雨和可能有害的或被污染的环境（因素的影响）。在这种情况下，地球上的地下城市和月球上或其他孤立环境中创造的基础没有什么不同。

Lesson 21

How Tunnels Are Built

After the general direction for a tunnel has been determined, the next steps are a geological survey of the site and a series of borings to obtain specific information on the strata through which the tunnel may pass①. The length and cross section of a tunnel generally are governed by the use for which it is intended②, but its shape must be designed to provide the best resistance to internal and external forces. Generally, a circular or nearly circular shape is chosen.

In very hard rock, excavation usually is accomplished by drilling and blasting. In soft to medium-hard rock, a tunnel-boring machine typically does the excavating work. In soft ground, excavation usually is accomplished by digging or by advancing a shield and squeezing the soft material into the tunnel. In all cases the excavated rock or earth, called muck, is collected and transported out of the tunnel. In underwater tunneling, a shield is used to advance the work. Another method of building an underwater tunnel is to sink tubular sections into a trench dug at the bottom of a river or other body of water.

1. Hard-rock tunnels

Short tunnels through hard rock are driven only from the portals, but longer ones usually are driven also from one or more intermediate shafts. Some long tunnels have been built with the aid of a small pilot tunnel driven parallel to the main tunnel and connected with it by crosscuts at intervals. The pilot tunnel not only furnishes additional points of access but also a route for removing muck and for ventilation ducts and drainage lines.

Another method is the heading-and-bench system, formerly used on most large tunnels because it required smaller amounts of powder and permitted simultaneous drilling and mucking (removal of excavated material). The upper portion of the tunnel is driven ahead of the lower part which is called the bench. A separate crew is thus able to muck in the lower portion of the tunnel while the upper portion is being drilled.

With improvements in tunneling methods and machinery, the full-face method of attack, previously used only in small tunnels, came into common use in building large ones. This change was partly brought about by the jumbo, a movable platform on which numerous rock drills are mounted. By this device, a large part of tunnel's face can be drilled at one time. The fullface method became the commonest and fastest way to drive a tunnel.

2. Soft-ground tunnels

Some tunnels are driven wholly or mostly through soft material. In very soft ground, little or no blasting is necessary because the material is easily excavated.

At first, forepoling was the only method for building tunnels through very soft ground. Forepoles are heavy planks about 5 feet (1.5 meters) long and sharpened to a point. They were inserted over the top horizontal bar of the bracing at the face of the tunnel. The forepoles were then driven into the ground of the face with an outward inclination. After all the roof poles were driven for about half of their length, a timber was laid across their exposed ends to counter any strain on the outer ends. The forepoles thus provided an extension of the tunnel support, and the face was extended under them. When the ends of the forepoles were reached, new timbering support was added, and the forepoles were driven into the ground for the next advance of the tunneling.

The use of compressed air simplified working in soft ground. An airlock was built, though which men and equipment passed, and sufficient air pressure was maintained at the tunnel face to hold the ground firm during excavation until timbering or other support was erected.

Another development was the use of hydraulically powered shields behind which cast-iron or steel plates were placed on the circumference of the tunnel. These plates provided sufficient support for the tunnel while the work proceeded, as well as full working space for men in the tunnel.

3. Underwater tunnels

The most difficult tunneling is that undertaken at considerable depths below a river or other body of water. In such cases, water seeps through porous material or crevices, subjecting the work in progress to the pressure of the water above the tunneling path. When the tunnel is driven through stiff clay, the flow of water may be small enough to be removed by pumping. In more porous ground, compressed air must be used to exclude water. The amount of air pressure that is needed increases as the depth of the tunnel increases below the surface.

A circular shield has proved to be most efficient in resisting the pressure of soft ground, so most shield-driven tunnels are circular. The shield once consisted of steel plates and angle supports, with a heavily braced diaphragm across its face. The diaphragm had a number of openings with doors so that workers could excavate material in front of the shield. In a further development, the shield was shoved forward into the silty material of a riverbed, thereby squeezing displaced material through the doors and into the tunnel, from which the muck was removed. The cylindrical shell of the shield may extend several feet in front of the diaphragm to provide a cutting edge. A rear section, called the tail, extends for several feet behind the body of the shield to protect workers. In large shields, an erector arm is used in the rear side of the shield to place the metal support segments along the circumference of the tunnel.

The pressure against the forward motion of a shield may exceed 1000 pounds per square foot (4880 kg/m^2). Hydraulic jacks are used to overcome this pressure and advance the shield, producing a pressure of about 5000 pounds per square foot (24500 kg/m^2) on the outside surface of the shield.

Shields can be steered by varying the thrust of the jacks from left side to right side or from top to bottom, thus varying the tunnel direction left or right or up or down. The jacks shove against the tunne llining for each forward shove. The cycle of operation is forward shove, line, muck, and then another forward shove. The shield used about 1955 on the third tube of the Lincoln Tunnel in New York City was 18 feet (5.5 meters) long and 31.5 feet (9.6 meters) in diameter. It was moved ahead 32 inches (81.2 centimeters) per shove, permitting the fabrication of a 32-inch support ring behind it.

Cast-iron segments commonly are used in working behind such a shield. They are erected and bolted together in a short time to provide strength and watertightness. In the third tube of the Lincoln Tunnel each segment is 7 feet (2 meters) long, 32 inches (81 centimeters) wide, and 14 inches (35.5 centimeters) thick, and weighs about 1.5 tons. These sections form a ring of 14 segments that are linked together by bolts. The bolts were tightened by hand and then by machine. Immediately after they were in place, the sections were sealed at the joints to ensure permanent watertightness.

4. Sunken-tube tunnels

Where the riverbed subsoil is firm and the river current is not excessive, shore-fabricated tunnel sections can be towed over a prepared trench in the river bottom and sunk into place to form an underwater tunnel. The first major prefabricated, floated, and sunken tunnel was the Detroit River Tunnel between Detroit and Windsor, Ontario. This vehicular tunnel was built in 1906—1910. The next important vehicular tunnel built by this method was the Posey Tube, which was completed in 1928. It runs under a saltwater arm between Oakland and Alameda, Calif. Since then many other sunken-tube tunnels have been built under rivers and saltwater bodies, notably the Transbay Tunnel between Oakland and San Francisco.

The cylindrical tunnel sections usually are made of steel in an onshore yard. Each section is about 300 feet (90 meters) long and 28 to 48 feet (8.5—14.6 meters) in diameter. After the openings at each end of a section are closed with steel bulkheads, the tube is ready for launching in the manner of a ship. Once in the water, a section is ballasted with concrete until a minimum buoyancy is attained. The section is towed to the tunnel site. Before the arrival of the section, dredges and underwater excavators dig a trench to the proper depth of the tunnel. When the tube section is positioned precisely over its final location, additional concrete is added until the section sinks into the prepared trench. All sections of the tunnel are transported and sunk in place in the same way.

Each section has projecting plates or flanges that fit over or into the preceding section in the manner of a male and female electrical connection. After on section and a succeeding one has been sunk, divers engage the flanges and tighten the bolts. Steel plates are slid down around the joint between the two closed bulkheads. The joint is then sealed with concrete to ensure watertight links between the sections. After all of the sections are placed and joined, they are covered over with fill to give them stability and protection. Thus the sunken-tube technique is an underwater version of the old cut-and-cover method.

In completing the work, crews enter from the portals at each end of the tunnel and cut away the steel bulkheads as they approach the center of the tube. Concrete then is placed for the interior lining of the tube, providing a good appearance and greater safety. Tiles, duct linings wiring, pumps, and piping are then added.

I. New Words

1. boring *vt.* 钻探，钻孔
2. strata *n.* 地层，岩层（stratum 的复数）
3. shaft *n.* 矿井，竖井
4. crosscut *n.* 横切，横巷，斜路
5. muck *n.* 腐殖土，出渣，挖除软土
6. bench *n.* 阶地，台阶，工作台
7. jumbo *n.* 凿岩台车，隧道盾构
8. blast *vi.* 爆炸，爆破
9. inclination *n.* 倾斜，倾角，斜度
10. circumference *n.* 周围，圆周，周边
11. porous *adj.* 多孔的，疏松的，有孔的
12. exclude *vt.* 隔绝，排除
13. shove *vt.* 挤，推，推动
14. cylindrical *adj.* 圆筒形的，圆柱体的
15. shield *n.* 盾构，护罩，掩护支架
16. steer *vt.* 操纵，调整，转向
17. jack *n.* 千斤顶，起重器
18. segment *n.* 部分，节，段
19. onshore *adj.* 向着海岸的，在岸上
20. dredge *n.* 挖掘，挖泥，疏浚
21. flange *n.* 凸缘，缘板，翼缘
22. bulkhead *n.* 隔壁，隔墙

II. Phrases and Expressions

1. medium-hard   中等硬度

2. hard-rock tunnel 硬岩隧道

3. pilot tunnel [隧道] 导洞

4. soft-ground tunnel 软土层隧道

5. cast-iron 铸铁的，坚固的

6. shield-driven 盾构法

7. sunken-tube tunnel 沉管隧道

8. cut-and-cover 随挖随填

9. seal with 用……密封

Ⅲ. Notes

① "through which…" 为介词＋关系代词引导的定语从句，这里 "which" 替代 "strata（地层）"，译为 "隧道可能穿越的地层"。

② "for which…" 也是介词引导的限定性定语从句，这里 "which" 替代 "use"。可见，为了将专业知识表述清楚严谨，大量使用定语从句是科技英语的一个重要特征，使得科技英语句子更加周密复杂。

Ⅳ. Exercises

1. Answer the following questions briefly.

(1) How many categories have tunnels been designed?

(2) How many approaches are introduced in the passage to construct a hard-rock tunneling?

(3) What are the procedures to construct a soft-ground tunneling?

2. Translate the following sentences into Chinese.

(1) Another method of building an underwater tunnel is to sink tubular sections into a trench dug at the bottom of a river or other body of water.

(2) Some long tunnels have been built with the aid of a small pilot tunnel driven parallel to the main tunnel and connected with it by crosscuts at intervals.

(3) A circular shield has proved to be most efficient in resisting the pressure of soft ground, so most shield-driven tunnels are circular.

(4) In large shields, an erector arm is used in the rear side of the shield to place the metal support segments along the circumference of the tunnel.

(5) Shields can be steered by varying the thrust of the jacks from left side to right side or from top to bottom, thus varying the tunnel direction left or right or up or down.

Ⅴ. Expanding

Know more about the shapes of highway tunnels.

There are three main shapes of highway tunnels—circular, rectangular, and horseshoe or curvilinear (Fig. 21-1 to Fig. 21-3). For example, rectangular tunnels are often constructed by either the cut and cover method, by the immersed method or by jacked box tunneling. Circular tunnels are generally constructed by using either Tunnel Boring Machine (TBM) or by drill and blast in rock. Horseshoe configuration tunnels are generally constructed by using drill and blast in rock or by the Sequential Excavation Method (SEM).

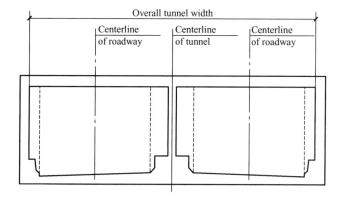

Fig. 21 - 1 Two-cell rectangular tunnel

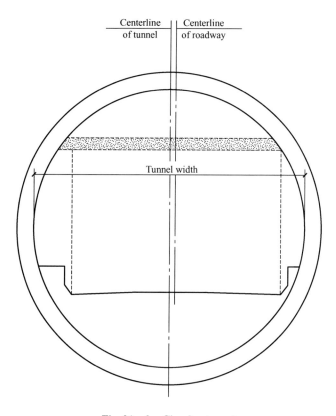

Fig. 21 - 2 Circular tunnel

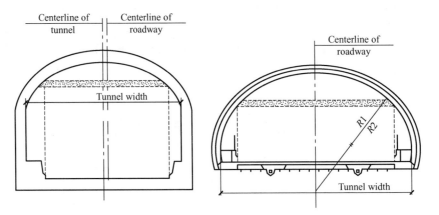

Fig. 21 – 3 Horseshoe and curvilinear (oval) tunnels

Ⅵ. Reading Material

Introduction to Tunnel Engineering

Almost every tunnel is a solution to a specific challenge or problem. In many cases, that challenge is an obstacle that a roadway or railway must bypass. They might be bodies of water, mountains or other transportation routes. Even cities, with little open space available for new construction, can be an obstacle that engineers must tunnel beneath to avoid. In the case of the Holland Tunnel, the challenge was an obsolete ferry system that strained to transport more than 20000 vehicles a day across the Hudson River. For New York City officials, the solution was clear: build an automobile tunnel under the river. The tunnel made an immediate impact. On the opening day alone, 51694 vehicles made the crossing, with an average trip time of just 8 minutes. And sometimes, tunnels offer a safer solution than other structures, which can be clearly shown in the case of the Seikan Tunnel in Japan. Tunnel engineering makes many vital underwater and underground facilities possible.

Technically, a tunnel is a horizontal passageway located underground. Three broad categories of tunnels are mining, public works and transportation. Mine tunnels are used during one extraction, enabling laborers or equipment to access mineral and metal deposits deep inside the earth. These tunnels are made using similar techniques as other types of tunnels, but they cost less to build. They are not as safe as tunnels designed for permanent occupation, however. Public works tunnels carry water, sewage or gas lines across great distance. By the 20th century, trains and cars had become the primary form of transportation, leading to the construction of bigger, longer tunnels to pass efficiently through an obstacle, such as a mountain. The Holland Tunnel, completed in 1927, was one of the first roadway tunnels and is still one of the world's greatest engineering projects. Fig. 21 – 4 is the Gotthard Base Tunnel, a railway tunnel under construction in Switzerland.

Fig. 21 – 4 The Gotthard Base Tunnel

A tunnel, at its most basic, is a tube hollowed through soil or stone. Constructing a tunnel, however, is one of the most complex challenges in the field of civil engineering. It is elaborate, but also important, to perform the tunnel type study as early as possible in the planning process and select the most suitable tunnel type for the particular project requirements. The selection, or to say, how a tunnel is built depends on the geometrical configurations, the ground conditions, and environmental requirements. For example, tunneling underwater demands a unique approach that would be impossible or impractical to implement above ground. The preliminary road tunnel type selection process is shown in Fig. 21 – 5.

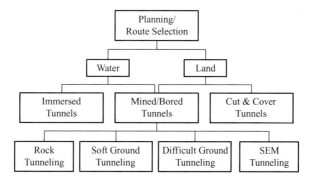

Fig. 21 – 5 Tunnel selection process

Shallow-depth tunnels, such as rapid-transit lines, underpasses, and end sections of tunnels through hills, are built by cut-and-cover method, generally used when the tunnel profile is shallow and the excavation from the surface is possible, economical, and acceptable. This method needs to excavate a trench from the surface, within which a concrete tunnel is constructed. With bottom-up construction, the completed tunnel is covered up, and the surface reinstated. With top-down construction, the walls are constructed first, the roof next, and then backfilled and the surface reinstated. Excavation and construction of the floors below roof level then follow.

Tunneling in rock today is primarily by drill-and-blast or by using a TBM (Tunnel Boring Machine). The basic approach of drill and blast excavation is to drill a pattern of small holes, load them with explosives, and then detonate those explosives thereby creating an opening in the rock. The blasted and broken rock (muck) is then removed and the rock surface is supported so that the whole process can be repeated as many times as necessary to advance the desired opening in the rock. Drill-and-blast tunnels can be any shape whereas most TBMs are only capable of drilling circular holes. TBMs excavate rock mass in a form of rotating and crushing by applying enormous pressure on the face with large thrust forces while rotating and chipping with a number of disc cutters mounted on the machine face as shown in the Fig. 21 – 6.

Fig. 21 – 6 Tunnel Boring Machine face with disc cutters for hard rock, Australia

Shield tunneling is the use of shield for tunnel excavation, lining and other operations of construction. Shield is special equipment with a revolving cutter wheel inside the guard. With soil immediately extruded, precast concrete segments are being erected in the shield tail. This method is not only characterized by high speed and constant quality of the tunnel body, but also has less impact on the surrounding building. Generally, it is used in non-cohesive, soft ground and indispensable for tunneling in loose sand, gravel, or silt and in all types of clay below the water table.

Sequential Excavation Method (SEM), also known as the New Austrian Tunneling Method (NATM), was developed in Austria but is now used worldwide. It is a tunneling method adapted to the excavation of variable and non-circular cross-section reaches of tunnel, such as highway ramps and subway stations. This underground method divides the space (cross-section) to be excavated into segments, then mines the segments sequentially, one portion at a time. By adjusting the construction sequence expressed mainly in round length, timing of support installation and type of support, it allows for tunneling through rock, soft ground and a variety of difficult ground conditions. The excavation can be carried out with common mining methods and equipment (often a backhoe), but as applied to soft ground tunneling, SEM generally cannot compete with tunneling machines for long running tunnels but often is a viable method for short tunnels.

Immersed tunnels are made from very large precast concrete or concrete-filled steel element that are fabricated in the dry, floated to the site, placed in a prepared trench below water, and connected to the previous elements, and then covered up with backfill. Immersed tunnels can be constructed in ground conditions that would make bored tunneling difficult or expensive, such as the soft alluvial deposits characteristic of large river estuaries, but when rock has to be excavated under water, immersed tunneling may be less cost effective.

Originally developed from pipe jacking technology, jacked box tunneling is a unique tunneling method, generally used in soft ground, for constructing shallow rectangular road tunnels beneath critical facilities (such as operating railways, major highways and airport runways), where TBM mining would not be economical or cut-and-cover methods would be too disruptive to overlying surface activities. Jacked box tunneling has mostly been used outside of United States (Taylor et al, 1998) until it was successfully applied to the construction of three short tunnels beneath a network of rail tracks at South Station in downtown Boston.

The box structure is constructed on jacking base in a jacking pit located adjacent to one side of an existing railway. A tunneling shield is provided at the front end of the box and hydraulic jacks are provided at the rear. The box is advanced by excavating ground from within the shield and jacking the box forward into the opening created at the tunnel heading. In order to maintain support to the tunnel face, excavation and jacking normally carried out alternately in small increments.

As tools improve, engineers continue to build longer and bigger tunnels. Recently, advanced technology promises to expedite excavation and ground support. The next generation of tunnel-boring machines will be able to cut 1600 tons of muck per hour. Engineers are also experimenting with other rock-cutting methods that take advantage of high-pressure water jets, lasers or ultrasonics. With new technologies and techniques, tunnels that seemed impossible even 10 years ago suddenly seem doable.

参 考 译 文

第 21 课　如何开挖隧道

在一条隧道的大致方向定下来之后，下一步就是调查隧道沿途地层的钻孔资料并获取具体的地层信息。隧道长度和横断面通常由其用途决定，但是其形状必须设计成对内外荷载产生最佳抗力的形式。通常会选择圆形或近似的圆形。

在非常坚硬的岩石中，通常采用钻机和爆破开挖。在软弱到中硬岩石，采用隧道挖掘机是典型的开挖方式。在软弱土层，通常采用盾构和挤压软弱土质的方式向前推进。在所有岩石或土层的开挖方式中，淤泥土要被收集起来运出隧道。在开挖水下隧道时，要采用盾构向前推进。另一种开挖水下隧道的方法是将深管放入河底或水中其他位置的已开挖的深沟中。

1. 硬岩隧道

穿过硬岩石短隧道仅从入口开挖，但是较长的隧道通常是从一个或几个地方同时开挖。有些长隧道是在平行于主隧道开挖的小型导洞辅助下建造的。导洞与主隧道之间每隔一段距离由横巷连通。导洞不仅是通道的附属设施，也是运土、通风、排水的通路。

另一种方法是采用正台阶开掘系统，以前被用于大型隧道，因为它仅要求更少的火药并且允许同时钻孔和运土（转移开挖材料）。上部隧道导向下部——这就叫做阶地，一个独立的施工队就可以在上部钻孔的同时在下部运土。

随着隧道开挖方法和机械设备的改进，以前仅用于小型隧道的全面施工方法，也开始普遍用于修建大型隧道。这种改变部分原因是隧道钻车——一种装有大量岩石钻头的可移动平台的引进。利用这种设备，一大片隧洞面可同时钻探。全面施工法已变成最普遍最迅速的开挖隧道方法。

2. 软土层隧道

有一些隧道是全部或绝大部分穿越软土层。在很软的土层中很少甚至不需要爆破，因为土质非常容易开挖。

一开始，超前伸梁掘进法是在软弱土层中建造隧道的唯一方法。超前掘进伸梁是一块大约 5 英尺（1.5m）长并且前端被锐化成一点的重重的厚板。它们被插入隧道表面的支撑柱的顶层水平条内。然后，超前伸梁向外倾斜钻入土层表面，在所有顶层杆被插入一半深度后，一根木料被交叉放置在它们的外露端来抵抗所有的外部应变。伸梁就这样提供了一种可以伸缩的坑道支撑，表面在其下伸出来。当杆的末端伸到后，再增设新的木支撑，伸梁被插入土中供隧道下一节使用。

压缩空气的使用简化了软土中的施工。首先建成一个空气锁，人和设备通过它进出，在开挖过程中，足够的空气压力来维持坑道表面的坚固直至木支撑或其他支撑竖立起来。

另一种发展是使用水压力盾构，其后在隧道四周嵌固铸铁或钢板支撑。这些钢板在施工过程中为隧道提供了足够的支撑力，同样也为施工人员提供了足够的施工空间。

3. 水下隧道

施工最困难的隧道是在一条河的下面一定深度处或水域其他部分中开挖隧道。在这种情况下，水会通过可渗透材料或裂缝渗出，影响了在上部水压力作用下的隧道施工进度。当隧道穿越黏稠土质时，水的流量也许会小到用水泵就可以抽干。在更大渗透性的土壤中，必须使用压缩空气来排水。所需空气压力随隧道距离表层的距离增加而增加。

实践证明，圆形盾构抵抗软土压力是最有效的，所以大多数盾构掘进的隧道都是圆形的。盾构曾经是由钢盘和角撑组成，再加上前端一个很重的支撑隔层。前端隔层有许多带门的开口以便工人可以在盾构前部开挖。在进一步的改进中，盾构向河床的粉质黏土中灌浆，通过这种方式将软弱土质从门里挤压进隧道，再从隧道将土运出。柱状的盾构壳可以在隔层前伸长数尺以提供一个剪切面。后面的部分，被称为尾巴，向后延伸数尺来保护施工工人。在大型的盾构机械中，一个起重臂被用在盾构的尾部沿隧道四周来代替金属支撑段。

盾构向前移动的阻力可能超过 1000 磅/平方英尺（4880kg/m^2）。液压千斤顶被用来克服这个阻力推动盾构向前，它能在盾构外侧提供 5000 磅/平方英尺的压力（24500kg/m^2）。

盾构可以通过操纵各种千斤顶使其能左右上下调整，以改变隧道左右上下的方向。千斤顶沿隧道线向每一段向前挤浆。整个循环过程是向前推进、定线、运土，然后另一段向前推进。1955年用于纽约市林肯隧道第三段的盾构长18英尺（5.5m），直径31.5英尺（9.6m）。每次向前推进32英寸（81.2cm），在其后制作一个32英寸的支撑环。

铸铁段通常被用在这样的盾构后面。它们被立起来并在短时间内栓接在一起以提供强度和防水。在林肯隧道的第三节，每段7英尺（2m）长，32英寸（81cm）宽，14英寸（35.5cm）厚，重大约1.5吨。这些段形成了由14个节段通过螺栓连接在一起的环。这些螺栓先由人工再用机械拧紧。一旦它们就位，这些节段在节点处密封以保证永久性防水。

4. 沉管隧道

当河床心土很坚固并且河水当前水量不是很充足的时候，岸边制造的隧道段可以拖拉到一个河床中已经准备好的壕沟里，并沉入水底以形成一个水下隧道。第一个主要的预制漂浮式沉管隧道是位于安大略省的底特律和温莎之间的底特律河隧道。这条运输隧道建于1906—1910年。第二条用类似方法建成的主要运输隧道是波西隧道，竣工于1928年。它位于加利福尼亚的奥克兰和阿拉米达之间的一条咸水河中。从那以后许多其他水底隧道被建于水下或是咸水区域中，特别是位于奥克兰和圣弗朗西思科的海湾隧道。

柱状隧道段通常位于河岸码头并由钢材制成。每段大约300英尺（90m）长，28~48英尺（8.5~14.6m）的直径。在每一段末端的开口处后面是用钢制挡水板封口，管子准备成发射船的样式。节段一旦放入水中，将使用混凝土压载直到达到最小浮力。然后这个节段被拖到隧道指定位置。在每一个节段就位之前，挖泥船和水下挖掘机会为隧道挖出一个合适深度的壕沟。当这个管段被精确地放到它最终就位的位置时，灌浆直至它沉到合适的沟槽里。隧道的所有分段以同样方式运输和下沉就位。

每段都有凸盘通过阴阳榫电焊与前段拼装。在每一段及其后续段下沉之后，潜水工人通过螺栓将凸盘紧固。钢盘被沿着两个封闭的挡水板之间的连接处滑下。接缝处再用混凝土封闭以增强段与段连接处的防水性。当所有的段被就位并连接好后，上面再填土以给予其稳定性和保护。可见，沉管技术是老式挖填土方法在水下的应用。

当施工结束时，工作人员从隧道每一节末端进入并在进入管道中心时移走钢制挡水板。然后再用混凝土做管道内衬，以获得更好的外观和更大的安全性。随后贴瓷砖，做风道衬里，安电线、水泵、管道等。

Lesson 22
Types of Bridges Ⅰ

A bridge is a structure providing passage over an obstacle such as a valley, road, railway, canal, river, without closing the way beneath. The required passage may be for road, railway, canal, pipeline, cycle track or pedestrians.

The branch of civil engineering which deals with the design, planning construction and maintenance of bridge is known as bridge engineering. Designs of bridges vary depending on the function of the bridge and the nature of the terrain where the bridge is constructed.

There are six main types of bridges: beam bridges, cantilever bridges, arch bridges, suspension bridges, cable-stayed bridges and truss bridges.

1. Beam bridges

Beam bridges are horizontal beams supported at each end by piers. The earliest beam bridges were simple logs that sat across streams and similar simple structures. In modern times, beam bridges are large box steel girder bridges. Weight on top of the beam pushes straight down on the piers at either end of the bridge. They are made up mostly of wood or metal. Fig. 22-1 shows a beam bridge style.

Fig. 22-1　Beam bridge

The beam bridge, also known as a girder bridge, is a firm structure that is the simplest of all the bridge shapes. Both strong and economical, it is a solid structure comprised of a horizontal beam, being supported at each end by piers that endure the weight of the bridge and the vehicular traffic. Compressive and tensile forces act on a beam bridge, due to which a strong beam is essential to resist bending and twisting because of the heavy loads on the bridge. When traffic moves on a beam bridge, the load applied on the beam is transferred to the piers. The top portion of the bridge, being under compression, is shortened, while the bottom portion, being under tension, is consequently stretched and lengthened. Trusses made of steel are used to support a beam, enabling dissipation of the compressive and tensile forces. In spite of the reinforcement by trusses, length is a limitation of a beam bridge due to the heavy bridge and truss weight. The span of a beam bridge is controlled by the beam size since the additional material used in tall beams can assist in the dissipation of tension and compression.

Extensive research is being conducted by several private enterprises and the state agencies to improve the construction techniques and materials used for the beam bridges. The beam bridge design is oriented towards the achievement of light, strong, and long-lasting materials like reformulated concrete with high performance characteristics, fiber reinforced composite materials, electro-chemical corrosion protection systems, and more precise study of materials. Modern beam bridges use prestressed concrete beams that combine the high tensile strength of steel and the superior compression properties of concrete, thus creating a strong and durable beam bridge. Box girders are being used that are better designed to undertake twisting forces, and can make the spans longer, which is otherwise a limitation of beam bridges. The modern technique of the finite element analysis is used to obtain a better beam bridge design, with a meticulous analysis of the stress distribution, and the twisting and bending forces that may cause failure.

2. Cantilever bridges

A cantilever bridge is a bridge built using cantilevers, structures that project horizontally into space, supported on only one end① (Fig. 22 - 2). For small footbridges, the cantilevers may be simple beams, however, large cantilever bridges designed to handle road or rail traffic use trusses built from structural steel, or box girders built from prestressed concrete. The steel truss cantilever bridge was a major engineering breakthrough when first put into practice, as it can span distances of over 460 meters, and can be more easily constructed at difficult crossings by virtue of using little or no falsework.

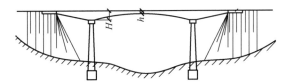

Fig. 22 - 2 **Cantilever bridge**

A simple cantilever span is formed by two cantilever arms extending from opposite sides of the obstacle to be crossed, meeting at the center. In a common variant, the suspended span, the cantilever arms do not meet in the center, instead, they support a central truss bridge which rests on the ends of the cantilever arms. The suspended span may be built off-site and lifted into place, or constructed in place using special traveling supports.

A common way to construct steel truss and prestressed concrete cantilever spans is to counterbalance each cantilever arm with another cantilever arm projecting the opposite direction, forming a balanced cantilever. When they attach to a solid foundation, the counterbalancing arms are called anchor arms. Thus, in a bridge built on two foundation piers, there are four cantilever arms: two which span the obstacle, and two anchor arms which extend away from the obstacle. Because of the need for more strength at the balanced cantilever's supports, the bridge superstructure often takes the form of towers above the foundation piers.

Steel truss cantilevers support loads by tension of the upper members and compression of the lower ones. Commonly, the structure distributes the tension via the anchor arms to the outermost supports, while the compression is carried to the foundations beneath the central towers. Many truss cantilever bridges use pinned joints and are therefore statically determinate with no members carrying mixed loads.

Prestressed concrete balanced cantilever bridges are often built using segmental construction. Some steel arch bridges are built using pure cantilever spans from each side, with neither false-work below nor temporary supporting towers and cables above. These are then joined with a pin, usually after forcing the union point apart, and when jacks are removed and the bridge decking is added the bridge becomes a truss arch bridge. Such unsupported construction is only possible where appropriate rock is available to support the tension in the upper chord of the span during construction, usually limiting this method to the spanning of narrow canyons.

3. Arch bridges

An arch bridge is a bridge with abutments at each end shaped as a curved arch (Fig. 22 - 3). Arch bridges work by transferring the weight of the bridge and its loads partially into a horizontal thrust restrained by the abutments at either side. A viaduct may be made from a series of arches, although other more economical structures are typically used today.

Fig. 22 - 3 Arch bridge

There are some variations of arch bridges:

1) Corbel arch bridges

The corbel arch bridge is a masonry or stone bridge where each successively higher course cantilevers slightly more than the previous course. The steps of the masonry may be trimmed to make the arch have a rounded shape. The corbel arch does not produce thrust, or outward pressure at the bottom of the arch, and is not considered a true arch. It is more stable than a true arch because it does not have this thrust. The disadvantage is that this type of arch is not suitable for large spans.

2) Aqueducts and canal viaducts

In some locations it is necessary to span a wide gap at a relatively high elevation, such as when a canal or water supply must span a valley. Rather than building extremely large arches, or very tall supporting columns, a series of arched structures are built one atop another, with wider structures at the base.② Roman civil engineers developed the design and constructed highly refined structures using only simple materials, equipment, and mathematics. This type is still used in canal viaducts and roadways as it has a pleasing shape, particularly when spanning water, as the reflections of the arches form a visual impression of circles or ellipses.

3) Deck arch bridges

This type of bridge comprises an arch where the deck is completely above the arch. The area between the arch and the deck is known as the spandrel. If the spandrel is solid, usually the case in a masonry or stone arch bridge, it is call a closed-spandrel arch bridge. If the deck is supported by a number of vertical columns rising from the arch, it is known as an open-spandrel arch bridge.

4) Through arch bridges

This type of bridge comprises an arch which supports the deck by means of suspension cables or tie bars. These through arch bridges are in contrast to suspension bridges which use the catenary in tension to which the aforementioned cables or tie bars are attached and suspended.

5) Tied arch bridges

Also known as a bowstring arch, this type of arch bridge incorporates a tie between two opposite ends of the arch. The tie is capable of withstanding the horizontal thrust forces which would normally be exerted on the abutments of an arch bridge.

Ⅰ. New Words

1. terrain *n.* 地形，地面，地域
2. cable *n.* 缆索，钢丝绳
3. stay *n.* 拉索，拉杆；*v.* 牵拉，支撑
4. pier *n.* （桥）墩，支柱，码头
5. log *n.* 原木，木材，木料
6. girder *n.* 主梁，纵梁，桁架
7. span *n.* 跨距，跨径
8. column *n.* 圆柱，柱形物
9. bending *n.* 弯曲，弯折，挠度
10. twisting *n.* 扭曲，翘曲
11. footbridge *n.* 人行桥
12. falsework *n.* 临时支架，脚手架
13. anchor *n.* & *v.* 锚固，锚碇，固定
14. jack *n.* 千斤顶，起重器
15. deck *v.* 作桥面；*n.* 桥面
16. canyon *n.* 峡谷
17. viaduct *n.* 高架（跨线）桥，栈道
18. masonry *n.* 砖石，砌体
19. spandrel *n.* 拱肩，拱腹

Ⅱ. Phrases and Expressions

1. beam bridge   梁式桥
2. cantilever bridge   悬臂桥
3. arch bridge   拱桥
4. suspension bridge   悬索桥，吊桥
5. cable-stayed bridge   斜拉桥
6. truss bridge   桁架桥
7. tensile strength   拉伸强度，抗拉强度
8. electro-chemical corrosion   电化学腐蚀
9. prestressed concrete   预应力混凝土
10. box girder   箱梁
11. finite element analysis   有限元分析
12. corbel arch bridge   叠涩拱桥
13. deck arch bridge   上承式拱桥
14. through arch bridge   下承式拱桥
15. suspension cable   悬索
16. tie bar   系杆
17. tied arch bridge   系杆拱桥
18. bowstring arch   系杆拱
19. thrust force   推力

Ⅲ. Notes

① 此句"is"后面的表语连用三个并列句，来解释什么是"cantilever bridge（悬臂桥）"。

② 句中的"rather than"为"是……而不是……"之意，即肯定 rather than 前面的内容，否定 rather than 后面的内容。

Ⅳ. Exercises

1. Translate the following sentences into Chinese.

(1) The branch of civil engineering which deals with the design, planning construction and maintenance of bridge is known as bridge engineering.

(2) A common way to construct steel truss and prestressed concrete cantilever spans is to counterbalance each cantilever arm with another cantilever arm projecting the opposite direction, forming a balanced cantilever, when they attach to a solid foundation, the counterbalancing arms are called anchor arms.

(3) These through arch bridges are in contrast to suspension bridges which use the catenary in tension to which the aforementioned cables or tie bars are attached and suspended.

2. Fill in the following blanks with correct names of the components in an arch bridge (Fig. 22 - 4).

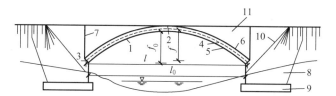

Fig. 22 - 4 Arch bridge segments

(1) _____ (2) _____ (3) _____ (4) _____ (5) _____
(6) _____ (7) _____ (8) _____ (9) _____ (10) _____
(11) _____

V. Expanding

Know about the chief influencing factors on designing a bridge.

The chief factors in deciding whether a bridge will be built as a girder, cantilever, truss, arch, suspension, or some other types are: (1) location, for example, across a river, (2) purposes, for example, a bridge for carrying motor vehicles, (3) span length, (4) strength of available materials, (5) cost, (6) beauty and harmony with the location.

VI. Reading Material

Construction of Concrete Bridges

At the stage of mixing, concrete behaves essentially as a fluid and must, therefore, be placed within confining formwork (Fig. 22 - 5) whilst it gradually hardens into the required structural shape. In this case the concrete has been transported from the mixer by crane.

Fig. 22 - 5 Placing and compaction of the concrete

There are essentially two quite distinct methods of construction:
(1) Placing the concrete directly into its final position (Fig. 22 - 6) in the structure (known as in situ concrete construction), concrete can often be transported from the mixer more conveniently by pumping it through pipes, the increased fluidity required for this

Fig. 22-6 Concrete pumping

being obtained by incorporating a Pozzolan cement.

(2) Placing the concrete into moulds (usually under factory conditions) from which the concrete member is subsequently removed, transported to site and erected in position (known as precast concrete construction).

Both systems have their advantages and disadvantages, and it is desirable that students begin to make their own observations on this.

For placing concrete in situ, not only is temporary formwork required but also supporting structures, termed falsework. The amount of such temporary work required can be considerable as shown in Fig. 22-7. Formwork for concrete will often require supporting during construction and in major bridges the temporary structures required can be considerable. This shows a concrete bridge under construction but steel girders support the formwork until the concrete has hardened, whilst additional steel girders facilitate the transfer of the formwork from one span to the next.

Fig. 22-8 shows an attractive concrete bridge in Switzerland built in 1929 (but one which poses a difficult constructional problem for in situ concrete). In this case the design of the supports to the formwork must involve as much structural engineering as the design of the final structure. In multi-storey structures it will be appreciated that in situ concrete floor construction involves an awkward transfer of forms and falsework from one floor to the next.

Fig. 22-7 Concrete bridge formwork   Fig. 22-8 A concrete arch bridge

The use of precast concrete members avoids some of these problems. Large span bridges can be constructed using precast segments of beam which are subsequently prestressed together. This may require falsework to support the segments whilst in situ concrete joints between the segments are made prior to the prestressing. Two views of a precast concrete segmental bridge under construction using false-work as temporary supports until the segments have been prestressed together. Fig. 22 - 9(a) shows the precast segments are temporarily stacked on site after delivery from the factory and prior to positioning on the temporary supports. In Fig. 22 - 9(b), units are positioned on the temporary supports. Note the positions of the ducts for the prestressing tendons. As the bridges are continuous over several spans, the prestressing tendons will need to be near the top of the units adjacent to a pier but near the bottom in units positioned near mid-span.

(a)　　　　　　　　　　　　　　　　　(b)

Fig. 22 - 9　Precast concrete segmental bridge

(a) Precast segments; (b) Precast segments in position

Alternatively, over difficult terrain, the precast units may be added one by one to give a cantilever form of construction which avoids the need for most of the temporary supports. Fig. 22 - 10 shows a precast concrete segmental bridge under construction using the cantilever method.

Fig. 22 - 10　Bridge built with cantilever method

参考译文

第22课 桥梁的类型Ⅰ

桥梁是提供通道跨越诸如山谷、道路、铁路、水道、河流等障碍物的结构,避免封闭下方路线。它为道路、铁路、水道、管道、自行车道或行人提供所需的通道。

桥梁工程被认为是土木工程的一个分支,它是处理桥梁的设计、建设规划和维护的部分。桥梁的设计根据桥梁的作用及桥梁建造地点的地形特点而改变。

桥梁主要有六种类型:梁式桥、悬臂桥、拱桥、悬索桥、斜拉桥和桁架桥。

1. 梁式桥

梁式桥是将水平梁的各端支撑在桥墩上。最早的梁桥是使用简单的原木跨越溪流和类似的简单结构物。在现代,梁式桥可以采用大型的钢箱梁结构。梁上的重量直接向下传递给桥梁每端的桥墩。它们主要使用木材或金属建造。如图22-1所示为一座梁式桥形式。

图22-1 梁式桥

梁式桥又称为板梁桥,它是一种在所有桥梁形状中最简单的坚固结构。它既牢固又经济,是一种由水平梁组成的实体结构,各端支撑在承担桥梁和车辆通行重量的桥墩上。由于强大的梁在桥梁重载的作用下必然承受弯曲和扭曲,这样梁桥上就作用有压力和拉力。当交通在梁桥上移动时,作用在梁上的荷载传递到桥墩上。桥梁的上部受压变短,而下部受到拉伸,因此在拉伸作用下变长。钢材制造的桁架被用来加固梁,使得拉力和压力分散。尽管采用桁架加固,由于承受沉重的桥梁和桁架重量,梁式桥的长度受到了限制。梁式桥的跨径受梁的尺寸限制,因为用于高梁的附加材料有助于耗散拉伸和压缩。

为了改善梁桥使用的施工技术和材料,在一些私人企业和国家机构的指导下进行了大量的研究。梁式桥的设计朝着获得轻型、坚固和耐久材料的方向发展,诸如具有高性能的新配方混凝土、纤维增强复合材料、电化学腐蚀防护系统以及更精确的研究材料。现代化的梁式桥使用预应力混凝土梁,它结合了钢材的高抗拉强度和混凝土的优秀的抗压特性,因此制造出坚固、耐久的梁桥。使用箱梁是因为其被优化设计可以用来承受扭力,而且其跨径更大,而大跨径则是梁式桥的一个限制因素。随着对应力分布和可能造成(结构)失效的扭力和弯曲力的细致分析,现代的有限元分析技术的应用可以获得更优的梁桥设计。

2. 悬臂桥

悬臂桥是使用悬臂建造的桥梁,结构水平地在空间伸出,只在一端支撑(图22-2)。对于小的人行桥,悬臂可以是简支梁,然而,大的为承受公路或铁路交通设计的悬臂桥使用结构钢制造的桁架修建,或者用预应力混凝土建造成箱梁。当钢桁架悬臂桥首次使用时,成为一项重要的工程突破,它能跨越460m的距离,使用很少或者不使用脚手架就可以在复杂的交叉口更方便地建造。

一个简支悬臂跨由两个伸出臂组成，伸出臂是由所跨越障碍的对面延伸出来，在跨中汇合。通常情况，悬臂跨的悬臂并不在跨中汇合，相反的，它们支撑一个搁置在悬臂端部的中央桁架桥。悬臂跨可以在场外建造并吊装到位，或者在现场使用特别的移动支架建造。

建造钢桁架和预应力混凝土悬臂跨的常见方法是用另一个朝相反方向伸出的悬臂来平衡每个悬臂，形成平衡悬臂，当其附属于实体基础时，平衡悬臂称为锚固臂。这样，在一座建于两个基础桥墩上的桥梁中会有四个悬臂：两个跨越障碍，两个朝远离障碍方向延伸的锚固臂。由于需要在平衡悬臂的支撑处具有更大的强度，因此桥梁上部结构在基础桥墩处经常采用塔的形式。

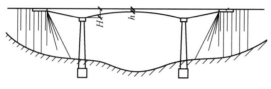

图 22 - 2　悬臂桥

钢桁架悬臂通过上部构件受拉和下部构件受压承担荷载。一般地，结构通过锚固臂将拉力分散到最外面的支撑处，而压力传递到中央塔下的基础。许多桁架悬臂桥使用铰接，这样的静定结构没有构件承担混合荷载。

预应力混凝土平衡的悬臂桥通常使用预制拼装施工方法建造。一些钢拱桥每端都使用纯悬臂跨修建，既没有在下面使用脚手架也没有在上部使用临时的支撑塔和拉索。通常在结合处将预制段分开后，再用销钉连接在一起，当千斤顶移除后增设桥面板，桥梁就成为一座桁架拱桥。这种无支撑施工方法只有在施工时桥跨上弦杆拉力可获得适当的岩石支撑的地方使用，因而经常限制了这种方法跨越狭窄的峡谷。

3. 拱桥

拱桥是每端使用墩台、形状如弯曲拱的一种桥梁（图 22 - 3）。拱桥受力是通过将其重量和荷载部分转化为水平推力，该力受到每端墩台的约束作用。尽管今天通常使用其他更经济的结构形式，但高架桥还可由一系列拱桥建成。

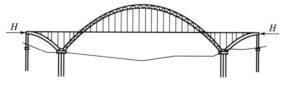

图 22 - 3　拱桥

拱桥可以变化为多种形式。

1）叠涩拱桥

叠涩拱桥是一种砌体或石材的桥梁，它的每一层都比前一层高并且悬臂稍长向内堆叠而成。砌体阶梯会被修整得使拱形成一个圆形。叠涩拱桥不会在拱的底部产生推力或向外的压力，因此它被认为不是一个真正的拱结构。由于它不具有推力，使得其比真正的拱更加稳定。这种类型的拱的缺点是不适合于大跨径。

2）渡槽和水渠高架桥

有些地点需要在相对高海拔地区跨越较宽的间隙，比如当一条水渠或供水管线必须跨越山谷的时候。一系列的拱形结构是一个叠着另一个地修建，在基础部位结构宽度较大，而不是建成特大跨度的拱或较高墩柱的结构。罗马的土木工程师发展了这种设计，并且仅使用简单的材料、装备和运算就修建了高度精炼的结构。这种类型仍然在水渠渡槽和公路中使用，因为其具有美观的外形，特别是在跨越水道的时候，拱形与倒影形成了圆形或椭圆形的视觉效果。

3）上承式拱桥

这种类型的桥梁由桥面完全在拱之上的拱结构组成。拱和桥面之间的部分称为拱肩。通常在砌体和石材拱桥的情况下，拱肩是实体的，它被称为实腹式拱桥。如果桥面板在拱上由一系列的垂直立柱支撑，则被称为空腹式拱桥。

4）下承式拱桥

这种类型的桥梁由通过吊索或系杆支撑桥面的拱结构组成。下承式拱桥与悬索桥相比，后者的悬链缆索在上述提到的吊索或系杆的连接和悬吊下处于受拉状态。

5）系杆拱桥

系杆拱又称为弓弦拱，这种类型的拱桥在拱的相对两端由系杆连接。系杆能够承受通常施加在拱桥墩台上的水平推力。

Lesson 23
Types of Bridges II

1. Suspension bridges

A suspension bridge is a type of bridge in which the deck is hung below suspension cables on vertical suspenders (Fig. 23 – 1). This type of bridge dates from the early 19th century, while bridges without vertical suspenders have a long history in many mountainous parts of the world.

Fig. 23 – 1 Suspension bridge

This type of bridge has cables suspended between towers, plus vertical suspender cables that carry the weight of the deck below, upon which traffic crosses. This arrangement allows the deck to be level or to arc upward for additional clearance. The suspension bridge is often constructed without falsework.

The suspension cables must be anchored at each end of the bridge, since any load applied to the bridge is transformed into a tension in these main cables. The main cables continue beyond the pillars to deck-level supports, and further continue to connections with anchors in the ground. The roadway is supported by vertical suspender cables or rods, called hangers. In some circumstances the towers may sit on a bluff or canyon edge where the road may proceed directly to the main span, otherwise the bridge will usually have two smaller spans, running between either pair of pillars and the highway, which may be supported by suspender cables or may use a truss bridge to make this connection. In the latter case there will be very little arc in the outboard main cables.

The main forces in a suspension bridge of any type are tension in the cables and compression in the pillars. <u>Since almost all the force on the pillars is vertically downwards and they are also stabilized by the main cables, the pillars can be made quite slender, as on the Severn Bridge[①], near Bristol, England.</u> In a suspended deck bridge, cables suspended via towers hold up the road deck. The weight is transferred by the cables to the towers, which in turn transfer the weight to the ground.

Assuming a negligible weight as compared to the weight of the deck and vehicles being supported, the main cables of a suspension bridge will form a parabola. One can see the shape from the constant increase of the gradient of the cable with linear distance, this increase in gradient at each connection with the deck providing a net upward support force. Combined with the relatively simple constraints placed upon the actual deck, this makes the suspension bridge much simpler to design and analyze than a cable-stayed bridge, where the deck is in compression.

The suspension bridge has some advantages over other bridge types:

(1) Longer main spans are more achievable than with any other types of bridges.

(2) Less material may be required than other bridge types, even at spans they can achieve, leading to a reduced construction cost.

(3) Except for installation of the initial temporary cables, little or no access from below is required during construction, for example allowing a waterway to remain open while the bridge is built above.

(4) May be better able to withstand earthquake movements than heavier and more rigid bridges can.

However, the suspension bridge has some disadvantages too compared with other bridge types:

(1) Considerable stiffness or aerodynamic profiling may be required to prevent the bridge deck vibrating under high winds.

(2) The relatively low deck stiffness compared to other (non-suspension) types of bridges makes it more difficult to carry heavy rail traffic where high concentrated live loads occur.

(3) Some access below may be required during construction, to lift the initial cables or to lift deck units. This access can often be avoided in cable-stayed bridge construction.

2. Cable-stayed bridges

A cable-stayed bridge is a bridge that consists of one or more columns, with cables supporting the bridge deck (Fig. 23 - 2). There are two major classes of cable-stayed bridges: In a harp design, the cables are made nearly parallel by attaching cables to various points on the tower so that the height of attachment of each cable on the tower is similar to the distance from the tower along the roadway to its lower attachment. In a fan design, the cables all connect to or pass over the top of the towers.②

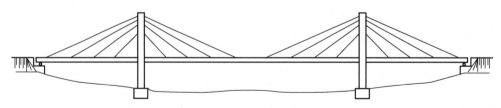

Fig. 23 - 2 Cable-stayed bridge

Compared with other bridge types, the cable-stayed is optimal for spans longer than typically seen in cantilever bridges and shorter than those typically requiring a suspension bridge. This is the range in which cantilever spans would rapidly grow heavier if they were lengthened, and in which suspension cabling does not get more economical were the span to be shortened.

A multiple-tower cable-stayed bridge may appear similar to a suspension bridge but in fact they are very different in principle and in the method of construction. In the suspension bridge, a large cable is made up by "spinning" small diameter wires between two towers, and at each end to anchorages into the ground or to a massive structure. These cables form the primary load-bearing structure for the bridge deck. Before the deck is installed, the cables are under tension from only their own weight. Smaller cables or rods are then suspended from the main cable, and used to support the load of the bridge deck, which is lifted in sections and attached to the suspender cables. As this is done the tension in the cables increases, as it does with the live load of vehicles or persons crossing the bridge. The tension on the cables must be transferred to the earth by the anchorages, which are sometimes difficult to construct due to poor soil conditions.

In the cable-stayed bridge, the towers form the primary load-bearing structure. A cantilever approach is often used for support of the bridge deck near the towers, but areas further from them are supported by cables running directly to the towers. This has the disadvantage, compared with the suspension bridge, the cables pull to the sides as opposed to directly up, requiring the bridge deck to be stronger to resist the resulting horizontal compression loads; but has the advantage of not requiring firm anchorages to resist a horizontal pull of the cables, as in the suspension bridge. All static horizontal forces are balanced so that the supporting tower does not tend to tilt or slide, needing only to resist such forces from the live loads.

Key advantages of the cable-stayed form are as follows:

(1) Much greater stiffness than the suspension bridge, so that deformations of the deck under live loads are reduced.

(2) Can be constructed by cantilevering out from the tower; and the cables act both as temporary and permanent supports to the bridge deck.

(3) For a symmetrical bridge, the horizontal forces balance and large ground anchorages are not required.

A further advantage of the cable-stayed bridge is that any number of towers may be used. This bridge form can be as easily built with a single tower, as with a pair of towers. However, a suspension bridge is usually built only with a pair of towers.

3. Truss bridges

A truss bridge is a bridge composed of connected elements which may be stressed from tension, compression, or sometimes both in response to dynamic loads (Fig. 23-3). Truss bridges are one of the oldest types of modern bridges. A truss bridge is economical to construct owing to its efficient use of materials.

Truss girders, lattice girders or open web girders are efficient and economical structural systems, since the members experience essentially axial forces and hence the

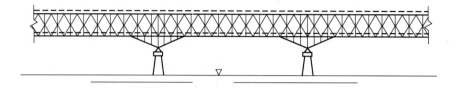

Fig. 23 – 3 Truss bridge

material is fully utilized. Members of the truss girder bridges can be classified as chord members and web members. Generally, the chord members resist overall bending moment in the form of direct tension and compression and web members carry the shear force in the form of direct tension or compression. Due to their efficiency, truss bridges are built over wide range of spans. Truss bridges compete against plate girders for shorter spans, against box girders for medium spans and cable-stayed bridges for long spans.

For short and medium spans it is economical to use parallel chord trusses such as Warren truss, Pratt truss, Howe truss, etc. to minimize fabrication and erection costs. Especially for shorter spans the Warren truss is more economical as it requires less material than either the Pratt or Howe trusses. However, for longer spans, a greater depth is required at the centre and variable depth trusses are adopted for economy. In case of truss bridges that are continuous over many supports, the depth of the truss is usually larger at the supports and smaller at mid-span.

Ⅰ. New Words

1. suspender *n.* 吊杆，悬杆（索）
2. clearance *n.* 净空，间距
3. pillar *n.* （索塔）柱
4. bluff *n.* 悬崖，陡岸
5. parabola *n.* 抛物线
6. stiffness *n.* 刚度，刚性
7. aerodynamic *adj.* 空气动力学的
8. anchorage *n.* 锚碇
9. web *n.* 腹板
10. erection *n.* 建造

Ⅱ. Phrases and Expressions

1. main cable   主缆
2. live load   活载，动载
3. hold up   举起，支撑
4. be similar to   与……相似，类似
5. in principle   大体上，原则上

6. multiple-tower cable-stayed bridge 多塔斜拉桥
7. horizontal force 水平力，横向力
8. dynamic load 动荷载，动力荷载
9. truss girder 桁架梁
10. lattice girder 井字梁
11. open web girder 空腹梁

Ⅲ. Notes

① "Severn Bridge"即英国的塞汶河桥，首创流线型箱梁桥面和混凝土桥塔，是一座主跨为988m的新型悬索桥。

② "In a harp design, ..."与"In a fan design, ..."分别译为"竖琴式索面设计……"与"扇形索面设计……"。扇形与竖琴形索面相比，索的利用率较高。但扇形布置斜拉索集中汇交于塔顶，塔顶构造细节较为复杂。反之，竖琴形索面由于所有斜拉索的斜角相同，塔上锚固点间距大，且所有斜拉索在梁端与塔端的锚固点结构细节相同，便于施工。

Ⅳ. Exercises

1. Answer the following questions.

（1）What are bridges and why are they necessary in transportation?

（2）How are bridges classified according to the text?

（3）What is the difference between pier and abutment?

（4）What are the variations of arch bridges?

（5）What are the advantages of cable-stayed bridges compared with suspension bridges?

2. Fill in the blanks with correct names of the components in a suspension bridge (Fig. 23 – 4).

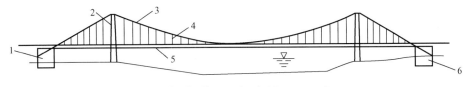

Fig. 23 – 4 Suspension bridge segments

（1）_____ （2）_____ （3）_____ （4）_____ （5）_____ （6）_____

Ⅴ. Expanding

Know about the construction materials and different forces in a bridge produced by imposed loads.

Bridge materials. The bridge designer can select from a number of modern high-strength materials, including concrete, steel, and a wide variety of corrosion-resistant alloy steels.

Bridge forces. A bridge must resist a complex combination of tension, compression, bending, shear and torsion forces. In addition, the structure must provide a safety factor as insurance against failure.

The forces that act on bridge structural members are produced by two kinds of loads—static and dynamic. The static load — the dead weight of the bridge structure itself — is usually the greatest load. The dynamic, or live load, has components, including vehicles carried by the bridge, wind forces, and accumulations of ice and snow.

Ⅵ. Reading Material

Techniques of Bridge Construction

The decision of how a bridge should be built depends mainly on local conditions. These include cost of material, available equipment, allowable construction time and environmental restriction. Since all these vary with location and time, the best construction technique for a given structure may also vary.

1. Incremental lauching or push-out method

In this form of construction the deck is pushed across the span with hydraulic rams or winches (Fig. 23-5). Decks of prestressed post-tensioned precast segments, steel or box girders have been erected. Usually spans are limited to 50—60 meters to avoid excessive deflection and cantilever stresses, although greater distances have been bridged by installing temporary support towers. Typically the method is most appropriate for long, multispan bridges in the range 300—600 meters, but, much shorter and longer bridges have been constructed. Unfortunately, this very economical mode of construction can only be applied when both the horizontal and vertical alignments of the deck are perfectly straight, or alternatively of constant radius. Where pushing involves a small downward grade (4%—5%) then a baking system should be installed to prevent the deck slipping away uncontrolled and heavy bracing is thus needed at the restraining piers.

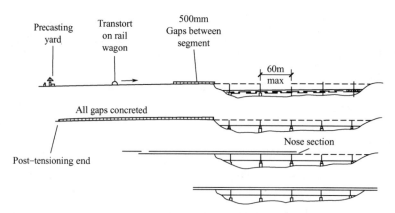

Fig. 23-5 Incremental lauching/push-out method

Bridge launching demands very careful surveying and setting out with continuous and precise checks made of deck deflections. A light aluminium or steel launching nose forms the head of the deck to provide guidance over the pier. Special teflon or chrome-nickel steel plate bearings are used to reduce sliding friction to about 5% of the weight, thus slender piers would normally be supplemented with braced columns to avoid cracking and other damage. These columns would generally also support the temporary friction bearings and help steer the nose.

In the case of precast construction, ideally segments should be cast on beds near the abutments and transferred by rail to the post-tensioning bed, the actual transport distance obviously being kept to the minimum. Usually a segment is cast against the face of the previously concreted unit to ensure a good fit when finally glued in placed with an epoxy resin. If this procedure is not adopted, gaps of approximately 500mm should be left between segments with the reinforcement running through and subsequently filled with concrete before post-tension begins. Generally all the segments are stressed together to form a complete unit, but when access or space on the embankment is at a premium it may be necessary to launch the deck intermittently to allow sections to be added progressively. The corresponding prestressing arrangements, both for the temporary and permanent conditions would be more complicated and careful calculations are needed at all positions.

The principal advantage of the bridge-launching technique is the saving in falsework, especially for high decks. Segments can also be fabricated or precast in a protected environment using highly productive equipment. For concrete segments, typically two segments are laid each week (usually 10—30mm in length and perhaps 300 to 400 tonnes in weight) and after post-tensioning incrementally launched at about 20 meters per day depending upon the winching/jacking equipment.

2. Balanced cantilever construction

Developments in box section and prestressed concrete led to short segments being assembled or cast in place on falsework to form a beam of full roadway width. Subsequently the method was refined virtually to eliminate the falsework by using a previously constructed section of the beam to provide the fixing for a subsequently cantilevered section. The principle is demonstrated step-by-step in the example shown in Fig. 23 - 6.

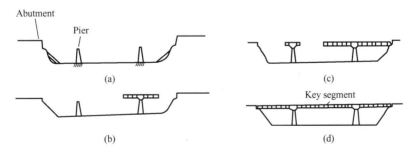

Fig. 23 - 6   Balanced cantilever construction

In the simple case illustrated, the bridge consists of three spans in the ratio 1 : 1 : 2. First the abutments and piers are constructed independently from the bridge superstructure. The segment immediately above each pier is then either cast in situ or placed as a precast unit. The deck is subsequently formed by adding sections symmetrically either side.

Ideally sections either side should be place simultaneously but this is usual impracticable and some imbalance will result from the extra segment weight, wind forced, construction plant and material. When the cantilever has reached both the abutment and centre span, work can begin from the other pier, and the remainder of the deck completed in a similar manner. Finally the two individual cantilevers are linked at the centre by a key segment to form a single span. The key is normally cast in situ.

The procedure initially requires the first sections above the column and perhaps one or two each side to be erected conventionally with either in situ or precast concrete and temporarily supported while steel tendons are threaded and post-tensioned. Subsequent pairs of sections are added and held in place by post-tensioning followed by grouting of the ducts. During this phase only the cantilever tendons in the upper flange and webs are tensioned. Continuity tendons are stressed after the key section has been cast in place. The final gap left between the two half spans should be wide enough to enable the jacking equipment to be inserted. When the individual cantilevers are completed and the key section inserted the continuity tendons are anchored symmetrically about the centre of the span and serve to resist superimposed loads, live loads, redistribution of dead loads and cantileverpretressing forces.

The earlier bridges were designed on the free cantilever principle with an expansion joint incorporated at the center. Unfortunately, settlements, deformations, concrete creep and prestress relaxation tended to produce deflections in each half span, disfiguring the general appearance of the bridge and causing discomfort to drivers. These effects coupled with the difficulties in designing a suitable joint led designers to choose a continuous connection, resulting in a more uniform distribution of the loads and reduced deflection. The natural movements were provided for at the bridge abutments using sliding bearings or in the case of long multi-span bridges, joints at about 500 meters centres.

参 考 译 文

第 23 课　桥梁的类型 Ⅱ

1. 悬索桥

悬索桥是桥面板通过竖直的吊索悬吊于悬索主缆之下的一种桥梁类型（图 23-1）。这种桥梁类型起源于 19 世纪初，在世界各地的许多山区采用无竖直吊索的此类桥梁有着悠久的历史。

这种类型的桥梁具有悬挂在桥塔之间的缆索，加上承担其下的用于通行交通的桥面板重量的竖直的吊索。这种布置要求桥面板是水平的或上拱以增加桥下净空。悬索桥通常采用无支架施工。

悬索桥的缆索必须锚固在桥梁的每端，这样，任何施加在桥梁上的荷载转化为主缆内

Lesson 23 Types of Bridges Ⅱ

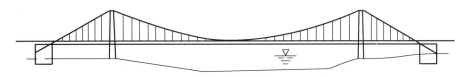

图 23 - 1 悬索桥

的拉力。塔柱顶部的主缆索延伸至桥面板标高处的支撑，然后继续与地面上的锚碇连接。路面由竖直的悬吊的索或杆（称为吊索）支撑。在一些情况中，桥塔坐落在绝壁或峡谷边缘，道路会直接延伸至主跨，否则，桥梁通常采用在每对塔柱和公路之间布置两个较小跨径，由吊索支撑或采用一座桁架桥来连接。在后一种情况中，外侧主缆会有很小的弧度。

任何类型的悬索桥的主要受力形式为缆索受拉与塔柱受压。由于几乎所有作用于塔柱上的力是竖直向下的，而且主缆使得塔柱稳定，因此塔柱可以建造得相当纤细，就像英国布里斯托尔的塞汶河桥一样。对于一座悬吊桥面板的桥梁，通过桥塔悬吊起来的缆索支撑起了路面板。其重量通过缆索传递给桥塔，依次再将重量传给地基。

假设与桥面板及其支撑的车辆重量相比主缆的重量微不足道，则悬索桥的主缆将形成抛物线形。可见随着直线距离的增加缆索梯度呈常量增量，在每个桥面板连接之间的梯度增量提供了向上的净支撑力。由于施加在实际桥面板上的约束相对简单，使悬索桥的设计和分析比斜拉桥更简单，它的桥面板是受压的。

悬索桥与其他桥型相比具有一些优点。

（1）与任何其他类型相比，可获得更长的主跨。

（2）与其他桥型相比需要更少的材料，即使它们的跨径一样，悬索桥可以减少建造费用。

（3）除了安装初始的临时缆索，在施工期不需要或几乎不需要下方通道，比如上部修建桥梁时允许水路保持开放。

（4）与更重和刚性更大的桥梁相比，可以更好地抗震。

然而，悬索桥与其他桥型相比也有一些缺点。

（1）需要相当大的刚度和满足空气动力学要求的外形来阻止桥面板在狂风下的振动。

（2）与其他（非悬吊）桥梁类型相比，相对较小的桥面刚度使其更难承受活荷载高集中的铁路重载交通。

（3）在施工过程中需要一些下方的通道，来提升最初的缆索或吊装桥面单元。这些通道在斜拉桥施工中通常可以避免。

2. 斜拉桥

斜拉桥是由一个或多个塔柱组成并通过缆索支撑桥面板的一种桥梁形式（图 23 - 2）。斜拉桥主要有两种：竖琴式索面设计，缆索布置为几乎平行的，通过在桥塔的不同点连接相应的拉索，这样每根拉索在塔上的连接高度和桥塔与沿着路面在低处的连接之间的距离是相似的；扇形索面设计，斜拉索都是在塔顶或穿过塔顶连接。

与其他桥型相比，斜拉桥最适于跨径长于悬臂桥的常用跨径并短于悬索桥通常所要求的跨径。在这一跨径范围内，悬臂跨径会随着长度加大而重量迅速增长，悬吊缆索即使缩短跨径也无法获得更大的经济性。

多塔斜拉桥可能看起来与悬索桥相似，但实际上它们在原理上和建造方法上是不同

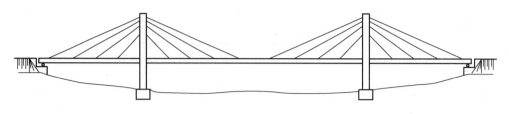

图 23-2 斜拉桥

的。对于悬索桥，一根粗缆索是在两个桥塔之间用小直径的钢丝通过"纺线"的方法建成，而且每端锚固到地基或巨型结构上。这些缆索形成了桥面板的最初的承重结构。在桥面安装前，缆索仅在其自身重量作用下承受拉力。然后，较小的吊索或杆从主缆悬吊下来，用于承担桥面板荷载，桥面板采用分段吊装并与悬挂的吊索连接。随着这些完成，车辆活载或人群通过桥梁时，缆索的拉力随之增加。缆索的拉力必须通过锚碇传给地基，因此在软弱地基条件下有时很难建造。

在斜拉桥中，桥塔形成最初的承重结构。在桥塔附近通常使用悬臂法支撑桥面板，但是更远的区域则由直接连接于桥塔的拉索支撑。与悬索桥相比，它具有的缺点是，与直接向上的方向相比斜拉索被拉到边上，需要桥面板具有足够大的强度以抵抗产生的水平压缩荷载；但是，优点是它不像悬索桥一样需要坚固的锚碇来抵抗缆索的水平拉力。所有静载产生的水平力都被平衡了，这样支撑桥塔就不会有倾斜和滑动的趋势，它只需抵抗由活载产生的此类水平力。

斜拉桥的主要优点如下。

（1）比悬索桥具有更大的刚度，这样在活载作用下桥面板的变形会减小。

（2）在桥塔部位可以采用悬臂法施工；斜拉索可以同时作为桥面板的临时和永久支撑。

（3）对于对称的桥梁，水平力相互平衡并且不需要大的岩土锚碇。

斜拉桥的更进一步的优点是可使用任何数量的桥塔。这种单塔的斜拉桥形式和双塔的一样容易建造。然而，悬索桥则通常只能建造一对桥塔。

3. 桁架桥

桁架桥是由承受拉力、压力或者有时在动载作用下同时承担拉压力的连接单元组成（图 23-3）。桁架桥是最古老的现代桥梁之一。由于有效地使用了材料，桁架桥施工经济。

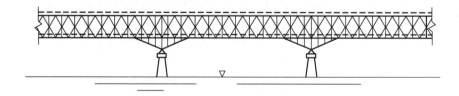

图 23-3 桁架桥

桁架梁、格构梁或者空腹式梁，由于构件主要承受轴向力，因此材料得到充分的利用，使得其形成高效和经济的结构系统。桁架桥的构件可以归类为弦杆和腹杆。一般情况下，弦杆抵抗由直接拉力和压力形成的全部的弯矩，而腹杆承受由直接拉力或压力形成的剪力。由于其高效性，桁架桥在较大跨径范围均有建造。桁架桥在较小跨径可与板梁桥竞争，在中等跨径可与箱梁桥竞争，并且在大跨径可与斜拉桥竞争。

在中小跨度中，使用平行弦桁架比较经济，例如华伦式桁架、普拉特桁架、豪威氏桁架等，以减少制造和安装费用。特别对于较小跨径，华伦式桁架比普拉特桁架和豪威氏桁架需要更少的材料，因此更经济。然而，对于大跨度情况，跨中需要较大的高度，为了经济性可采用变高度桁架。对于支撑在许多支座上的连续桁架桥的情况，桁架的高度通常在支点较大而在跨中较小。

Lesson 24
Hydrology and Hydraulics

1. Introduction

The objectives of a hydraulic design are to identify the stream forces that may cause harm to the bridge or roadway system and to provide a safe level of service acceptable to the needs of the traveling public without unreasonable effect on adjacent property.

1) Policy and coordination

Consideration of the effects of constructing a bridge across a waterway is the key to assuring the long-term stability of the structure. Confining the floodwater may cause excessive backwater or overtopping of the roadway, or it may induce excessive scour. These effects may result in damage to upstream land or endanger the bridge. Conversely, an excessively long bridge may cause flooding downstream or cost far more than that can be justified by the benefits obtained. Somewhere between these extremes is the design that will be the most economical to the public over a long period of time. The designer must evaluate existing upstream conditions as well as future upstream development—25 years, if possible—in sizing a structure. The amount of information available to assist the designer in predicting future development is limited.

2) Design responsibilities

Responsibilities for drainage design are divided between the bridge design section and the project development sections based primarily on the size of the drainage area. Bridge design is responsible for all watersheds over 300 acres (120 hectares) and locations where the existing structure opening exceeds 20ft.^{2} (1.86m^2). Project development is responsible for watersheds smaller than 300 acres (120 hectares).

The bridge design section is responsible for design of all locations requiring structures exceeding 20ft.^{2} (1.86m^2) of waterway opening for a single location. Bridge design will also be responsible for design of pipe culverts, closed drainage systems, roadside ditches, and stormwater management on "bridge only" projects.

If the designer is using a design flood frequency less than that the specifications requirement[①], a risk analysis must be performed. This needs approval by the bridge design engineer. Where a risk analysis is needed, a complete hydraulic report should be prepared giving consideration to each alternative under study. The lower level of study, risk assessment, should always be considered as the first course of action. Only if a detailed economic accounting of the risks and potential harm is needed should an extensive risk analysis be performed②. A detailed analysis must be performed in situations involving substantial losses resulting in high encroachment costs. It will always be necessary to apply good engineering judgment in determining the level of evaluation to be performed.

2. Hydrology and hydraulics

Hydrologic analysis is used to determine the rate of flow, runoff, or discharge that the drainage facility will be required to accommodate.

The design of highway facilities should be adequately documented. Frequently, it is necessary to refer to plans, specifications, and hydrologic analyses long after the actual construction has been completed. One of the primary reasons for documentation is to evaluate the hydraulic performance of structures after large floods to determine whether the structures performed as anticipated or to establish the cause of unexpected behavior. In the event of failure, it is essential that contributing factors be identified to avoid recurring damage.

The documentation of a hydrologic analysis is the compilation and preservation of all pertinent information on which the hydrologic decision was based. This might include drainage areas and other maps, field-survey information, source references, photographs, hydrologic calculations, flood-frequency analyses, stage-discharge data and flood history, including narratives from highway maintenance personnel and local residents who witnessed or had knowledge of an unusual event. Hydrologic data shown on project plans ensures a permanency of record, serves as a reference in making plan reviews, and aids field engineers during construction.

1) Estimating flood runoff and magnitudes

In Delaware, there are six methods of estimating flood magnitudes in sizing a waterway opening for a given structure: USGS Method, TR-55, Recorded Data, Published Reports, Rational Method, and TR-20.

USGS Method: DelDOT uses the equations in the current version of the US Geological Survey (USGS) publication *Technique for Estimating Magnitude and Frequency of Floods* in Delaware to estimate flood runoff for drainage areas of 300 acres (120 hectares) or greater. These equations are based on specific studies of the watersheds in Delaware and adjacent states. This method relies on data from stream flow gaging station records combined statistically within a hydrological homogenous region to produce flood-frequency relationships applicable throughout the region. If the designer is using gaging station records and wishes to evaluate these values for upstream or downstream sites, the procedures in the USGS publication should be followed.

From the study, it was concluded that reasonable estimates of flood runoff can be made by dividing the State into two regions. In the northern region, only the size of the drainage area, basin development factor, and storage are considered in the equations; the other factors that affect runoff are considered in the constants and exponents. In the southern region, basin relief, forest cover, and two soil categories must be considered in addition to the drainage area and storage.

2) Design flood frequency and frequency mixing

The design frequencies for bridges and pipe culverts for each highway functional classification are provided in the specifications. If a design frequency is less than the standard, the design must be based on a risk analysis and must be approved by the bridge design engineer.

Often, the designer is faced with the situation wherein the hydraulic characteristics of the subject facility are influenced by a flood condition of a separate and independent drainage course. For example, a small stream may outfall into a major river that itself is an outfall for a large and independently active watershed. It can reasonably be expected that these two waterways would seldom peak at the same time. Consequently, there are two independent events: one, a storm event occurring on the small stream; the other, a storm event applicable to the larger watershed.

For a given area ratio and design year, the designer should estimate the total flood discharge by adding the estimated discharges from the main stream and tributary based on the design year frequency. Two estimates must be made: one with the design storm applied to the main stream watershed and one with the tributary drainage area experiencing the design storm. The design should be based on the highest discharge.

The effects of tidal flows must be considered when the designer is evaluating the frequency mixing relationships.

3) Ice and debris

The quantity and size of ice and debris carried by a stream should be investigated and recorded for use in the design of drainage structures. The times of occurrence of ice or debris in relation to the occurrence of flood peaks should be determined, and the effect of backwater from ice or debris jams or recorded flood heights should be considered in using stream-flow records. The location of the constriction or other obstacle-causing jams, whether at the site or structure under study or downstream, should be investigated, and the feasibility of correcting the problem should be considered.

Under normal circumstances, one foot (0.3m) freeboard is sufficient to permit passage of ice flow and debris. At locations where large pieces or quantities of debris are anticipated, the designer should consider increasing the freeboard.

4) Tidal hydraulics bridges and culverts

At this time, there is no single authoritative reference for guiding the engineer in modeling the effect of tidal flows on the hydraulics of a structure. There are several circumstances in which the potential for tidal impacts is significant.

The size of the bridge opening may be controlled in a case of incoming (flood) tidal flows and peak storm discharge. Another consideration that may control the size of the opening is the storm surge at peak flood tidal flows. In the same manner, scour of the stream bottom is a concern on outgoing (ebb) tidal flows and peak storm discharges.

These and other combinations of tidal and storm flows must be considered in the sizing and design of a structure.

Ⅰ. New Words

1. hydrology *n.* 水文学
2. hydraulics *n.* 水力学
3. scour *vt.* 冲刷，疏通
4. watershed *n.* 分水岭，流域，集水区
5. encroachment *n.* 侵犯，超出界限
6. runoff *n.* 流走部分
7. discharge *vt.* 流出，排出
8. narrative *n.* 叙述，讲述
9. permanency *n.* 耐久性，永久
10. magnitude *n.* 量级，大小，重要性
11. homogenous *adj.* 结构相同的，同源发生的
12. outfall *n.* 河口，湖口，排水口
13. tributary *adj.* 支流的；*n.* 支流
14. tidal *adj.* 潮汐的，受潮汐影响的
15. debris *n.* 残骸，碎片
16. downstream *adv.* 顺流地，下游；*n.* 下移，下行

Ⅱ. Phrases and Expressions

1. drainage design　排水设计
2. flood frequency　洪水频率
3. design frequency　设计频率
4. tidal hydraulics　潮汐水力学

Ⅲ. Notes

① 句中的"design flood frequency"译为"设计洪水频率"，"specifications"指行业规范。

② "Only if"引导条件状语从句时，主句一般采用倒装句，因此主句将情态动词 should 提前，并使用了虚拟语气，表示在"Only if"假设条件下提出的建议。

Ⅳ. Exercises

1. Fill in the blanks with the following phrases to identify the typical river cross section (Fig. 24 - 1).

low water level　　　benchland　　　river channel　　　flood level
medium water level　　marginal bank　　main river channel

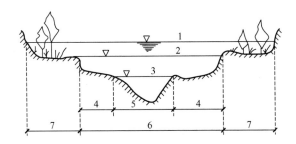

Fig. 24-1 River cross section

(1) _____ (2) _____ (3) _____ (4) _____ (5) _____
(6) _____ (7) _____

2. Translate the following sentences into Chinese.

(1) Consideration of the effects of constructing a bridge across a waterway is key to assuring the long-term stability of the structure.

(2) Only if a detailed economic accounting of the risks and potential harm is needed should an extensive risk analysis be performed.

(3) Hydrologic analysis is used to determine the rate of flow, runoff, or discharge that the drainage facility will be required to accommodate.

(4) The times of occurrence of ice or debris in relation to the occurrence of flood peaks should be determined, and the effect of backwater from ice or debris jams or recorded flood heights should be considered in using stream-flow records.

V. Expanding

Remember the following phrases and expressions which are important in hydrology and hydraulics.

1. alluvial fan   冲击扇
2. channel storage   槽蓄梁，槽蓄作用
3. flood routing   洪水演算
4. fluvial process   河床演变
5. gauging station   水文站
6. hydraulic geometry   水力几何学
7. ice-jam   冰塞
8. pilot channel   导河
9. river valley   河谷
10. sediment carrying capacity   挟沙能力
11. sediment transport   泥沙输送

Ⅵ. Reading Material

Flow around Bridge Piers and through Bridge Waterways

A common hydraulic problem encountered by engineers is that of flow through a bridge. Although the procedures for the structural design of bridges are well established, those relating to hydraulic design are relatively vague. Additionally, scour around piers and abutment causes many problems. Scour is the erosion of the river bed by flowing water. Particular difficulties are encountered in predicting the depth of scour, and in providing adequate scour protection. Floods, scour and movement of the foundations are the most common cause of bridge failure. In any particular year, hundreds of bridges around the world may be destroyed. Many involve loss of life when vehicles drive or fall off the collapsed bridge into the river. Fortunately this was not the case when part of the viaduct in Fig. 24-2 collapsed, probably as a result of scour. In February 1989 part of the railway viaduct over the River Ness collapsed, probably as a result of scour.

Fig. 24-2 Railway viaduct over the River Ness

When a bridge is built across a river, for reasons of economy it is frequently necessary to have piers and/or abutments in the channel. These obstruct the normal flow of water and cause an increase in upstream water level called the afflux. The variation of the afflux with upstream distance is the backwater curve. The greater the obstruction, the greater the afflux. Thus the smaller the distance between the abutments and the larger the number of piers, the greater the afflux and the risk of upstream flooding and flood damage.

The afflux is basically caused by energy head loss as the flow passes through the bridge opening. This loss arises from the need for the water on the upstream floodplain to contract through the opening and then, crucially, to expend back onto the floodplain. The energy head loss in expanding flow is generally quite large. To overcome this head loss and to maintain continuity if flow there has to be an increase in upstream water level (i. e. the afflux) to force the water through the obstruction. Usually the greatest afflux occurs between one and two spans upstream, that is between b and $2b$ where b is the opening width.

The afflux increases rapidly if the deck of the bridge comes into contact with the water, causing the bridge waterway opening to become submerged so that the flow resembles that through an orifice (free or submerged depending upon whether or not the downstream water level is also above the top of the opening). The expansion is now three dimensional and friction losses, which are normally relatively small, increase significantly. This is illustrated by Fig. 24-3 which shows a typical stage-discharge curve for a submerged bridge opening. The curve is initially concave downwards since in channel flow $Q \propto D^{5/3}$. The lower D_N curve shows the head-discharge relationship of an open channel, while the upper curve shows a typical relationship for a bridge located in the channel. The vertical difference between the two lines represents the afflux (or backwater), which increases rapidly after the waterway opening of height Y becomes submerged at $H_1/Y \geqslant 1.1$. The data are obtained from a 250-mm wide by 125-mm high rectangular bridge opening placed in a laboratory channel at a slope of 1/200. After submergence, in orifice flow, it is concave upwards since $Q \propto D^{1/2}$ (with the area of the opening constant).

It might be imagined that it would be possible to avoid the opening becoming submerged, but this is not always possible either because of cost or site conditions. For example, in Fig. 24-4 the bridge deck and railway line could not be raised because of the proximity to Exeter Station, consequently the deck and abutments have been rounded and slim piers constructed to optimise hydraulic efficiency. It is a trapezoidal flood relief channel for the River Exe at Exeter. The channel capacity is $450 m^3/s$. The sides of the channel are grassed while the base has a concrete lining. This adds to the difficulty of estimating the n value of the channel, and hence its capacity.

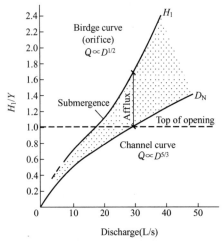

Fig. 24-3 Stage-discharge curve

Fig. 24-4 Flood relief channel

参 考 译 文

第 24 课　水文学和水力学

1. 导言

一项水力设计的目的就是确定可能引起桥梁或道路系统破坏的水流作用力，来根据旅行者的需求，提供一个他们可以接受的、安全的运营水平，而且不会对周围产业造成不合理的影响。

1) 政策和协调

考虑在水道上建造一座桥梁的影响因素是保证结构长期稳定的关键。洪水可以引起大量的壅水或漫过路面，并且引起过度冲刷。这些影响将导致上游陆地的损坏并且进一步危及桥梁。相反地，过长的桥梁会导致下游洪水泛滥并且可以证明成本比获得的收益要大得多。设计要介于这两个极端之间，在很长一段时间内对于公众是最经济的。设计者必须评估现有的河流上游条件，同时评估上游条件的发展——如果可能 25 年——由此确定结构的尺寸。帮助设计者预测未来发展状况的有效信息量是有限的。

2) 设计职责

根据排水面积的大小，排水设施设计的责任分为桥梁设计部门和项目发展部门。桥梁设计部门负责超过 300 英亩（120 公顷）的水域以及现有的结构开口处超过 20 平方英尺（1.86 平方米）的位置。项目发展部门负责 300 英亩（120 公顷）以内的水域。

桥梁设计部门需要对全部单处构筑物水路开口超过 20 平方英尺（1.86 平方米）的地方进行设计。桥梁设计部门还要负责设计管式涵洞、封闭的排水系统、路旁的沟渠以及仅应对暴雨设计的桥梁工程。

如果设计者使用的设计洪水频率小于规范要求，还必须要进行风险分析。这需要桥梁设计工程师的批准。在需要进行风险分析的地方，在研究中对所有可选择的事项进行考虑，准备出完整的水力报告。较低水平的研究，风险估价应该总是作为首先考虑的步骤。只有需要详细的经济方面的风险和潜在危害的报告书时，才进行广泛的风险分析。在实际的损失造成高昂的成本的情况下，必须进行详细的分析。用良好的工程判断力来决定应使用的评估水平总是必要的。

2. 水文学和水力学

使用水力分析可以确定排水设施需要容纳的水流速率、径流量，或者排水量。

进行公路设施的设计时需要有足够的文件支持。通常，在实际建筑已经全部完成后很长时间，查阅平面图、说明书以及对水文方面的分析仍是很有必要的。查阅文件的一个主要原因是在大的洪水过后评估结构的水文性能，来判断结构的工作是否像我们所预期的那样，或者确定产生没有预料到的状况的原因。在失败的事件中，重要的是辨别对失败起作用的要素，以避免破坏再次发生。

水文分析文件是对水文决策所依据的相关信息进行编辑和保存。这包括排水区域和其他区域的地图、实地测量信息、原始参考资料、照片、水文方面计算、洪水-频率分析、泄洪阶段的数据以及洪水历史，包括公路维护人员和当地目击了洪水或了解不寻常事件的当地居民的叙述。在工程设计图上说明的水文数据应确保记录的永久性，它作为对设计图进行回顾时的参考，并且在施工中可以辅助工地上的工程师。

1) 评估洪水径流及流量

在特拉华有六种方法，通过测定给定结构开放的排水沟的大小来评估洪水流量：USGS方法、TR-55、已记录的数据、出版的报告、推理的方法以及TR-20。

USGS方法：特拉华运输部使用的是美国地质调查（USGS）出版物《洪水流量及频率估算技术》中的公式，来评估特拉华洪水的大小和频率，用以估计排水面积为300英亩（120公顷）或更大的洪水径流量。这些公式基于对特拉华和相邻州的流域的详细研究上。这个方法依赖于河流观测站的记录数据，该数据是对水文方面同源的区域的统计组合，以提出可以应用到全部区域上的洪水频率关系。如果设计者使用河流观测站的记录并希望估计出上游和下游站的某点的值，则可以采用USGS出版物中的步骤。

从研究中可以推断，合理估计洪水径流可以通过将该州分为两个区域获得。在北部区域，在公式中只考虑了排水面积的大小、河槽发展因素及储水量；其他影响径流的因素被考虑到常数或指数中。在南方地区除了排水面积和储水量外，河槽地貌、森林覆盖，以及两种土壤类型均应被考虑。

2) 设计洪水频率和频率混叠

规范中列出了适合每种公路功能分类的桥梁和管式涵洞的设计频率。如果使用的设计频率小于规范要求，则设计必须给出风险分析，并且必须得到桥梁设计工程师批准。

设计者经常会遇到这样的情况：附属设施的水力特性受到分开或独立排水阶段时洪水情况的影响。例如，一个小的溪流会排入一条主干河流，而主干河流本身也是一个较大的独立活动的流域的出水口。可以合理地认为这两条水路很少会同时达到最大量。因此，有两个独立的事件：一个是暴雨发生在小溪流中；另一个是暴雨发生在较大的流域中。

对于一个给定的面积比率和设计年限，设计者应该评估总的洪水流量，基于设计年限频率把主要河流的估计流量加上附属流量。必须做出两方面的评估：一是将设计暴雨用于主要河流流域，一是将设计暴雨用于附属排水区域。设计应依据最大的排水量。

当设计者评估有混合关系的频率时，必须考虑潮汐流的影响。

3) 冰及杂物残骸

对一条河流携带的冰和杂物残骸的数量和大小应该进行调查并且记录下来，以备在设计排水结构时使用。应该确定冰和杂物残骸出现的次数与洪水峰值发生时间之间的关系，在使用水文记录时，应该考虑由于冰或杂物残骸的堵塞造成的壅水或有记录的洪水高度的影响。河道收缩或其他障碍物引起阻塞的位置，无论是在被研究结构物的所在地或是在其下游，均应被查明，并且应该考虑改正问题的可行性。

在正常的环境下，一英尺（0.3m）的出水高度就足以使冰流和垃圾碎片通过。在可预料到有垃圾碎片比较大或数量比较多的地方，设计者应该考虑将出水高度加大。

4) 桥梁和涵洞的潮汐水力学

到目前为止，还没有一个权威的参考书可以指导工程师来模拟结构在水力方面受潮汐流的影响。在一些情况下受到潮汐影响的可能性是值得注意的。

潮汐流（洪水）到来时以及暴雨高峰时的泄水量对桥梁通水孔的大小起控制作用。另一个需要考虑的控制桥梁通水孔大小的因素是洪水潮汐高峰时的暴雨水位。同样的，要考虑到河流底部受潮汐流流出（退下）以及暴雨泄洪高峰的冲刷。在确定构筑物的尺寸和设计时，这些问题以及潮汐流和暴雨流的结合问题均应被考虑。

Lesson 25

Road Engineering

1. A historical note

The first road builders of any significance in Western Europe were the Romans, to whom the ability to move quickly from one part of the Empire to another was important for military and civil reasons. Roman roads are characterized by their linearity and, in popular perception, by their durability. A good alignment was sought since this provided the most direct route and the risks of ambush in hostile territory were reduced. It was for this reason that the surface of the road was often elevated one meter or more above the local ground level to provide a clear view of the surrounding country, hence the modern term "highway". The durability of such pavements is less absolute but nevertheless well exceeds anything achieved for many centuries after the fall of the Empire.

A typical major Roman road in the UK consisted of several layers of material, increasing in strength from the bottom layers, perhaps of rubble, through intermediate layers of lime-bound concrete to an upper layer of flags or stone slabs grouted in lime. The total thickness of such a pavement would be varied according to the ground conditions. In sound ground a thickness approaching one meter might be found; elsewhere this would be increased as necessary.

During the Dark Ages — and indeed well after that — no serious attempt was made in the UK to either maintain or replace the Roman road network, which consequently deteriorated. By the end of the Middle Ages there was in practice no road system in the country. Such routes as existed were unpaved tracks, swampy and impassable for most of the year and dusty and impassable for the remainder. Diversions around particularly poor lengths of road, private land or difficult topography had resulted in sinuous alignments. The general lawlessness combined with these characteristics may discourage all but the most determined travelers.

The first small change in this state of affairs was bought about by an Act of 1555 which imposed a duty on each parish to maintain its roads and to provide a surveyor of highways. As this post was unpaid and under-resourced, and as the technical skills did not exist to match the task in hand, the obvious expectation that the post of surveyor was unpopular and ineffective was generally correct.

The lack of resources remained a problem for over a century. In the latter part of the 17th century the first experimental lengths of turnpike road were established on the Great

North Road (now the A1 Trunk Road). Turnpiking is a toll system whereby travelers pay for the use of the road. ① In the first part of the next century the parliament produced a series of Acts which enabled the establishment of turnpike trusts on main routes throughout the country. In this improved financial climate road building techniques were gradually developed through the work of such pioneers as Metcalf, Telford and the eponymous Macadam. By the 1830s a system of well-paved roads had evolved of such quality that they imposed little or no constraint on road traffic. Journey times were limited not by the state of the road but by the nature of road vehicles.

The next improvement in the speed and cost of travel came about as a result of a radical change in vehicle technology—the building of the railways. The effect of this was to reduce road traffic between towns to such a low level that the turnpike system became uneconomic. Although road building in towns continued, the turnpike trusts collapsed. Legislation in the late 19th century set the scene for the current administrative arrangements for highway construction and maintenance but the technology remained empirical and essentially primitive. Only in recent years has that situation changed to any great extent. ②

2. The aims of highway engineering

In order that economic activity can take place, people, goods and materials must move from place to place. The necessary movement has, to some extent, always been possible, but the growth in economic activity which characterized the Industrial Revolution in the 18th century of England and which has occurred or is occurring throughout the world since then, placed demands on the transport system which in its original primitive form it was quite unable to meet. This system developed to meet the new needs much more rapidly than it had previously, and the economy expended further, generating more traffic, and in this interactive way canals and turnpike roads were produced, then railways and most latterly a network of modern roads.

The tendency is for economic growth to be concentrated in areas where transport facilities are good. For example, the construction in the UK of a motorway network during the quarter century starting in the 1960s has increased access from formerly remote areas to the capital and to international links, and those areas have prospered. In the previous century the railways had a similar effect: areas formerly several days travel from any centers of population were, with the opening of a connecting railway, suddenly only a few hours away, and benefited as a result. Roads provide a key element of the infrastructure whose function is to promote economic activity and improve the standard of living of the population. Highway engineering is concerned with the best use of resources to ensure that a suitable network is provided to satisfy this need of an economically sophisticated society.

Originally, roads were little more than tracks across the countryside and were hard, dry and dusty in summer, and sodden and impassable in winter. The practice arose, initially in towns, of paving the surface of the road with resilient naturally occurring materials such as stone flags, and such a surface became known as a pavement. Today this term is applied to any surface intended for traffic and where the native soil has been protected from the harmful effects of that traffic by providing an overlay of imported or treated materials. The purpose of proving the protection is to enable traffic to move easily—and therefore more cheaply or quickly—along the road.

3. Highway types

1) Freeway

A freeway, as defined by statute, is a highway in respect to which the owners of abutting lands have no right or easement of access to or from their abutting lands, or in respect to which such owners have only limited or restricted right or easement of access. This statutory definition also includes expressways. The engineering definitions for use in this manual are:

(1) Freeway. A divided arterial highway with full control of access and with grade separations at intersections.

(2) Expressway. An arterial highway with at least partial control of access, which may or may not be divided or have grade separations at intersections.

2) Controlled access highway

In situations where it has been determined advisable by the Director or the CTC, a facility may be designated a "controlled access highway" in lieu of the designation "freeway". All statutory provisions pertaining to freeways and expressways are applied to controlled access highways.

3) Conventional highway

A highway without control of access may or may not be divided. Grade separations at intersections or access control may be used when justified at spot locations.

4) Highway

(1) Arterial highway. A general term denoting a highway primarily for through traffic usually on a continuous route.

(2) Bypass. An arterial highway that permits traffic to avoid part or all of an urban area.

(3) Divided highway. A highway with separated roadbeds for traffic in opposing directions.

(4) Major street or major highway. An arterial highway with intersections at grade and direct access to abutting property and on which geometric design and traffic control measures are used to expedite the safe movement through traffic.

(5) Radial highway. An arterial highway leading to or from an urban center.

(6) Through street or through highway. Every highway or portion thereof at the entrance to which vehicular traffic from intersecting highways is regulated by stop signs or traffic control signals or is controlled when entering on a separate right—turn roadway by a yield—right—of way sign.

5) Parkway

An arterial highway for noncommercial traffic, with full or partial control of access, and usually located within a park or a ribbon of park —like development.

6) Scenic highway

An officially designated portion of the State Highway System traversing areas of outstanding scenic beauty which together with the adjacent scenic corridors requires special scenic conservation treatment.

7) Street or road

(1) Cul-de-Sac street. A local street open at one end only, with special provisions for turning around.

(2) Dead end street. A local street open at one end only, without special provisions for turning around.

(3) Frontage street or road. A local street or road auxiliary to and located on the side of an arterial highway for service to abutting property and adjacent areas and for control of access.

(4) Local street or local road. A street or road primarily for access to residence, business, or other abutting property.

(5) Toll road, bridge or tunnel. A highway, bridge, or tunnel opening to traffic only upon payment of a direct toll or fee.

Ⅰ. New Words

1. significance *n.* 重要性，紧要，重大
2. ambush *n.* 伏击，埋伏
3. hostile *adj.* 敌人的，敌方的
4. rubble *n.* 毛石，块石
5. flag *n.* 薄层，薄层砂岩
6. swampy *adj.* 沼泽的，潮湿的，湿而松软的
7. diversion *n.* 绕行，绕路
8. topography *n.* 地形（测量），地形
9. sinuous *adj.* 曲折的，错综复杂的
10. act *n.* 决议，法令
11. parish *n.* 教区
12. turnpike *n.* 收费高速公路
13. macadam *n.* 碎石路面
14. prosper *vt.* 使兴隆，使繁荣，使成功
15. infrastructure *n.* 基础设施
16. sophisticated *adj.* 老练的，复杂的
17. sodden *adj.* 浸透了的，泡胀了的
18. resilient *adj.* 有弹性的
19. easement *n.* 缓和，减轻，使用权
20. designate *vt.* 指明，指派，称作
21. statutory *adj.* 有关法令的
22. pertain *vi.* 与……有关
23. roadbed *n.* 路床

24. expedite *vt.* 加快，促进

25. auxiliary *adj.* 辅助的，补充的

Ⅱ. Phrases and Expressions

1. lime-bound concrete　石灰土，石灰混凝土
2. in lieu of　代，代替
3. Cul-de-Sac　［法］死路
4. CTC（Centralized Traffic Control）　交通控制中心

Ⅲ. Notes

① "whereby" 为关系副词，可用 "by which" 替代，译为 "凭借，由此"。
② "Only" 置于句首，修饰强调作状语的 "in recent years"，主句采用部分倒装。

Ⅳ. Exercises

1. Answer the following questions briefly.

(1) What is the significance of highway engineering?

(2) What are the aims of highway engineering?

(3) How can highways be generally classified?

2. Fill in the blanks with the following phrases to identify the typical road distresses (Fig. 25-1).

Joint destruction　　　　Potholes　　　　Longitudinal and transverse cracking
Subgrade depression induced pavement cracking　　Corner break

Fig. 25-1　Road distresses

Ⅴ. Expanding

Remember the following phrases which are important in road engineering.
1. freeway 高速公路
2. classified highway 等级公路
3. arterial highway 干线公路
4. feeder highway 支线公路
5. national trunk highway 国家干线公路（国道）
6. provincial trunk highway 省干线公路（省道）
7. county road 县公路（县道）
8. expressway （城市）快速路
9. relief road 辅道
10. climatic zoning for highway 公路自然区划
11. highway right-of-way 公路用地
12. radial highway 辐射式公路
13. ring highway 环形公路
14. highway route 公路路线
15. highway alignment 公路线形
16. horizontal alignment 平面线形
17. vertical alignment 纵面线形
18. alignment elements 线形要素
19. technical standard of road 道路技术标准
20. boundary line of road construction 道路建筑限界

Ⅵ. Reading Material

Pavement Types

1. Introduction

Hard surfaced pavements, which make up about 60 percent of U. S. roads and 70 percent of Washington State roads are typically categorized into flexible and rigid pavement:

Flexible pavements are those which are surfaced with bituminous (or asphalt) materials. These types of pavement are called "flexible" since the total pavement structure "bends" or "deflects" due to traffic loads. A flexible pavement structure is generally composed of several layers of materials which can accommodate this "flexing".

Rigid pavements are those which are surfaced with Portland Cement Concrete (PCC). These types of pavements are called "rigid" because they are substantially stiffer than flexible pavements due to PCC's high stiffness.

Each of these pavement types distributes load over the subgrade in a different fashion.

Rigid pavement, because of PCC's high stiffness, tends to distribute the load over a relatively wide area of subgrade (Fig. 25 – 2). The concrete slab itself supplies most of a rigid pavement's structural capacity. Flexible pavement uses more flexible surface course and distributes loads over a smaller area. It relies on a combination of layers for transmitting load to the subgrade (Fig. 25 – 3).

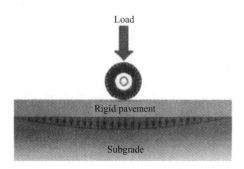

Fig. 25 – 2　Rigid pavement load distribution　　　Fig. 25 – 3　Flexible pavement load distribution

In general, both flexible and rigid pavements can be designed for long life (e. g., in excess of 30 years) with only minimal maintenance. Both types have been used for just about every classification of road. Certainly there are many different reasons for choosing one type of pavement or the other, some practical, some economical, and some political. As a point of fact, 93 percent of U. S. paved roads and about 87 percent of Washington State paved roads are surfaced with bituminous (asphalt) materials.

2. Flexible pavement

Flexible pavements are so named because the total pavement structure deflects, or flexes, under loading. A flexible pavement structure is typically composed of several layers of materials. Each layer receives the loads from the above layer, spreads them out, and then passes on these loads to the next layer below. Thus, the further down in the pavement structure a particular layer is, the less load (in terms of force per area) it must carry.

In order to take maximum advantage of this property, material layers are usually arranged in order of descending load bearing capacity with the highest load bearing capacity material (and most expensive) on the top and the lowest load bearing capacity material (and least expensive) on the bottom. The typical flexible pavement structure consists of:

(1) Surface course. This is the top layer and the layer that comes in contact with traffic. It may be composed of one or several different HMA sublayers.

(2) Base course. This is the layer directly below the HMA layer and generally consists of aggregate (either stabilized or unstabilized).

(3) Subbase course. This is the layer (or layers) under the base layer. A subbase is not always needed.

3. Rigid pavement

Rigid pavements are so named because the pavement structure deflects very little under loading due to the high modulus of elasticity of their surface course. A rigid pavement structure is typically composed of a PCC surface course built on top of either the subgrade or an underlying base course. Because of its relative rigidity, the pavement structure distributes loads over a wide area with only one, or at most two, structural layers.

The typical rigid pavement structure consists of:

(1) Surface course. This is the top layer, which consists of the PCC slab.

(2) Base course. This is the layer directly below the PCC layer and generally consists of aggregate or stabilized subgrade.

(3) Subbase course. This is the layer (or layers) under the base layer. A subbase is not always needed and therefore may often be omitted.

参 考 译 文

第 25 课　道 路 工 程

1. 历史记录

在西欧有着某种意义的第一条道路的建设者是罗马人，基于军事和民事的原因，由帝国的一个部分迅速转移到另一个部分的能力对罗马人来说非常重要。罗马道路的特点是它们是线性的，通俗来说，其具有耐久性。因为要提供最直达的道路，所以要寻求良好的线路线形，这样在敌对领土中遭遇伏击的风险才能降低。正是由于这个原因，路面通常比当地的地面标高高一米或更多，以提供清晰的视野来观察周围的情况，因此才出现了现代术语"高速公路"。这种路面的耐久性不是很绝对，但仍然远远超过帝国沦陷之后的许多世纪内所取得的任何成就。

在英国，典型的主要罗马道路由几层材料组成，从底层开始强度逐渐增加，底层可能由碎石构成，中间层由石灰结合混凝土构成，上层由石灰灌浆的石板构成。这种路面的总厚度将根据地面条件而变化。在稳定的地面上，路面厚度可能接近一米；其他地方的路面厚度必要时可以增大。

在黑暗时代（欧洲中世纪），并且从那之后在英国没有认真尝试维护或替代罗马公路网络，因此罗马公路网络恶化。到中世纪结束时，实际上该国还没有道路系统。现有的公路都是未铺装的小路，沼泽遍布，几乎整年都无法通行，其余部分也是尘土飞扬且难以通行。围绕在特别差的道路、私人土地或困难地形中的岔道弯曲地排列着。常见的违章乱建与这些特点结合起来，使所有人除了最坚定的旅行者外都望而却步。

这种情况的第一个小改变是通过制定1555法案来完成的，该法案规定每个教区都有责任维护道路，并有责任提供公路调查员。由于这个职位没有报酬且资源不足，而且由于技术技能不足以满足手头的任务，故很显然调查员的职位是不受欢迎的和无效的，尽管愿望是好的。

资源的匮乏在一个多世纪以来依然是个问题。在17世纪后半叶，在大北路（现在的

A1 干线路）上建立了第一个收费公路的试行路段。收费高速公路有一个收费系统，旅行者可凭借该系统支付公路使用费。在接下来的一个世纪的前叶，议会制定了一系列法令，就是在全国各主要路线上建立收费信托关卡。在这种改进的金融环境下，道路建造技术通过像麦特卡夫、泰尔福特和与其齐名的马卡达姆等开拓者的贡献而得到逐渐发展。到19世纪30年代，铺好的道路网络已经达到一定的质量，以至于它们对道路交通的羁绊很少甚至是没有。旅行时间不再受道路的状态所限，而是受车辆性质的限制。

接下来，旅行的速度和成本的改进是由于车辆制造技术的根本性变化——铁路建设的结果。这样做的效果是将城镇之间的道路交通降到一个很低的水平，使得收费系统变得不经济。虽然城镇道路建设会继续进行，但收费信托关卡就会失败。19世纪末期的立法为当前公路建设和维护的行政划分奠定了基础，但该技术仍然是经验性的，基本上是原始的。仅是近年来，这种情况才有了较大程度的改变。

2. 公路工程的目标

为了从事经济活动，人、物资和材料必须从一个地方迁移到另一个地方。必要的运动在某种程度上总是可能的，但是经济活动的增长对运输系统提出了要求，其原始的简单形式远远无法满足经济发展的需要，其特征是18世纪英格兰的工业革命，自那时以来该工业革命在世界各地发生或正在发生着。这个系统的开发是为了满足比以前发展更快的新需求，此时经济消耗更多，需要发展更多的交通，在这种互相影响下就兴起了运河和收费公路，然后是铁路，近来是现代公路网络。

有个趋势是经济增长集中的地区交通设施也好。例如，在20世纪60年代开始的四分之一个世纪期间，英国机动车道路网络的建设就增加了很多线路，如从偏远地区到首都及与国际公路网的接入，使得这些地区繁荣起来。在20世纪，铁路也有类似的发展效果：随着连接铁路的开通，过去距人口中心区有几天路程的地区，突然变成只有几小时的路程，因而大大受益。道路是基础设施的关键要素，其职能是促进经济活动和提高人民的生活水平。公路工程关注资源的最佳利用，确保提供合适的公路网，以满足成熟经济社会的这种需要。

最初，道路不过是横贯乡村的小道，在夏天很硬、很干燥且尘土飞扬，而在冬天，却又是泥泞不堪，难以通行。最初是在城镇开始修建道路，使用能产生自然弹性的材料来铺路的面层，如石板，这样的面层被称为碎石路面。今天，这一术语适用于任何与交通相关的路面，以及需要用进口的或处理过的材料做面层来保护当地土壤免受交通有害影响的地方。证明此种保护的目的在于使道路上的交通更顺畅——因此交通也就更便宜或更快速。

3. 公路类型

1）高速公路

根据法规定义的高速公路是一条公路，与其毗邻的土地的所有者没有权利或地役权进出这些毗邻的土地，或者这些拥有者只有限的或受约束的权利或地役权接近该片土地。这个法定定义也包括快速干道。本手册中使用的工程定义如下：

（1）高速公路。一条有分隔带的干线高速公路，可完全控制出入口，并在道路交叉口有立体交叉。

Lesson 25 Road Engineering

（2）快速干道。至少带有部分出入口控制的干线高速公路，其可以有也可以没有分隔带或者在道路交叉口有立体交叉。

2）出入管制高速公路

在已确定被主管或交通控制中心采纳的情况下，可以将设施定名为"出入管制高速公路"，以代替"高速公路"的命名。所有与高速公路和快速干道有关的法律规定都适用于出入管制高速公路。

3）常规公路

没有出入管制的高速公路可以有也可以没有分隔带。在交叉路口处的立体交叉或出入控制可以在合理的地方使用。

4）公路

（1）主干路。一个通用术语，表示通常是在一个连续的路线上的主要用于通过交通的高速公路。

（2）支线。要求交通避让部分或全部市区交通的主干路。

（3）分隔式公路。在相反的方向上有分离车行道承载交通的公路。

（4）主要街道或主要公路。主干路带有立体交叉，可直达邻接区域，并采取几何设计和交通控制措施以促进交通安全运行。

（5）辐射式公路。通往或离开城市中心的主干道。

（6）直通街道或直通公路。在每条公路或其入口部分，来自相交公路的车辆交通通过停车标志或交通控制信号灯来进行调控，或者当通过有优先权的右转路标时进入单独的右转弯道路时受到控制。

5）公园道路

用于非商业交通的主干道，全部或部分控制进出，并且通常位于公园内或类似公园的开发带。

6）景观公路

官方指定的穿越优秀景区的国家公路系统的一部分，与相邻的景区走廊一起需要景区特殊的保护治理。

7）街道或道路

（1）独头巷道。当地街道只有一端开放，有特殊的掉头规定。

（2）死路街道。当地街道只有一端开放，对掉头没有特殊的规定。

（3）临街街道或道路。位于主干路一侧的本地街道或辅助道路，用于服务临街建筑物和邻近区域以及用于控制出入。

（4）当地街道或当地道路。主要用于进入住宅、商业或其他邻接物业的街道或道路。

（5）收费公路、桥梁或隧道。公路、桥梁或隧道只有在直接付费或收费后才能开通。

Lesson 26
Traffic Engineering[1]

Traffic engineering is a branch of civil engineering that uses engineering techniques to achieve the safe and efficient movement of people and goods on roadways. It focuses mainly on research for safe and efficient traffic flow, such as road geometry, sidewalks and crosswalks, cycling infrastructure, traffic signs, road surface markings and traffic lights. Traffic engineering deals with the functional part of transportation system, except the infrastructures provided.

Traffic engineering is closely associated with other disciplines:
(1) Transport engineering.
(2) Pavement engineering.
(3) Bicycle transportation engineering.
(4) Highway engineering.
(5) Transportation planning.
(6) Urban planning.
(7) Human factors engineering.

Typical traffic engineering projects involve designing traffic control device installations and modifications, including traffic signals, signs, and pavement markings. However, traffic engineers also consider traffic safety by investigating locations with high crash rates and developing countermeasures to reduce crashes. Fig. 26-1 shows the complex intersections with multiple vehicle lanes, bike lanes, and crosswalks are common examples of traffic engineering projects. Traffic flow management can be short-term (preparing construction traffic control plans, including detour plans for pedestrian and vehicular traffic) or long-term (estimating the impacts of proposed commercial developments on traffic patterns). Increasingly, traffic problems are being addressed by developing systems for intelligent transportation systems, often in conjunction with other engineering disciplines, such as computer engineering and electrical engineering.

1. Traffic systems

Traditionally, road improvements have consisted mainly of building additional infrastructure. However, dynamic elements are now being introduced into road traffic management. Dynamic elements have long been used in rail transport. These include sensors to measure traffic flows and automatic, interconnected, guidance systems to manage traffic (for example, traffic signs which open a lane in different directions depending on the time of

Fig. 26 – 1 Complex intersection

day). Also, traffic flow and speed sensors are used to detect problems and alert operators, so that the cause of the congestion can be determined, and measures can be taken to minimize delays. These systems are collectively called intelligent transportation systems.

2. Lane flow equation

The relationship between lane flow (Q, vehicles per hour), maximum speed (v, kilometers per hour) and density (K, vehicles per kilometer) is,

$$Q = Kv$$

Observation on limited access facilities suggests that up to a maximum flow, speed does not decline while density increases. However, above a critical threshold, increased density reduces speed. Additionally, beyond a further threshold, increased density reduces flow as well.

Therefore, speeds and lane flows at bottlenecks can be kept high during peak periods by managing traffic density using devices that limit the rate at which vehicles can enter the highway (Fig. 26 – 2). Ramp meters, signals on entrance ramps that control the rate at which vehicles are allowed to enter the mainline facility, provide this function (at the expense of increased delay for those waiting at the ramps).

Fig. 26 – 2 A ramp meter limits the rate at which vehicles can enter the freeway

3. Highway safety

Highway safety engineering is a branch of traffic engineering that deals with reducing the frequency and severity of crashes. It uses physics and vehicle dynamics, as well as road user psychology and human factors engineering, to reduce the influence of factors that contribute to crashes.

A typical traffic safety investigation follows these steps:

(1) Identify and prioritize investigation locations. Locations are selected by looking for sites with higher than average crash rates, and to address citizen complaints.

(2) Gather data. This includes obtaining police reports of crashes, observing road user behavior, and collecting information on traffic signs, road surface markings, traffic lights and road geometry.

(3) Analyze data. Look for collisions patterns or road conditions that may be contributing to the problem.

(4) Identify possible countermeasures to reduce the severity or frequency of crashes.

① Evaluate cost/benefit ratios of the alternatives.

② Consider whether a proposed improvement will solve the problem, or cause "crash migration". For example, preventing left turns at one intersection may eliminate left turn crashes at that location, only to increase them a block away.

③ Are any disadvantages of proposed improvements likely to be worse than the problem you are trying to solve?

(5) Implement improvements.

(6) Evaluate results. Usually, this occurs some time after the implementation. Have the severity and frequency of crashes been reduced to an acceptable level? If not, return to step(2).

Ⅰ. New Words

1. sidewalk *n.* 人行道
2. crosswalk *n.* 人行横道
3. infrastructure *n.* 基础结构，基础设施
4. modification *n.* 修改，改变，改造
5. countermeasure *n.* 对策，反措施
6. detour *n.* 迂（回）路，便道，绕道
7. interconnect *vt.* 使互相连接，使互相联系
8. congestion *n.* 拥挤，挤满，超负荷
9. bottleneck *n.* 瓶颈，瓶颈口
10. ramp *n.* 斜面，斜坡，坡道
11. prioritize *vt.* 依主次程序排列，指定优先权
12. collision *n.* 碰撞，撞击
13. implement *n.* 工具，器具，用具

Ⅱ. Phrases and Expressions

1. traffic flow　交通流量
2. traffic sign　交通标志
3. road surface marking　路面标记
4. traffic light　交通灯
5. urban planning　城镇规划
6. crash rate　事故率
7. short-term　短期的
8. long-term　长期的
9. intelligent transportation system　智能交通系统
10. in conjunction with　连同，共同，与……协力
11. critical threshold　临界阈限，临界门槛
12. at the expense of　以……为代价，由……支付费用

Ⅲ. Notes

① 本篇文章内容比较简单，但是具有明显的科技英语的特点：在词汇上采用了大量科技英语词汇，即专业技术词汇；在句法层面上，大量使用了名词化结构、被动结构、非谓语动词短语、无人称句等；在修辞特点上，陈述句居多，依式行文，文笔朴素，结构严谨，逻辑性强，专业术语性强。

Ⅳ. Exercises

1. Answer the following questions briefly.

(1) What disciplines are traffic engineering associated with?

(2) What factors does the lane flow equation involve?

(3) What steps does a typical traffic safety investigation include?

2. Translate the following sentences into Chinese.

(1) It focuses mainly on research for safe and efficient traffic flow, such as road geometry, sidewalks and crosswalks, cycling infrastructure, traffic signs, road surface markings and traffic lights.

(2) Traditionally, road improvements have consisted mainly of building additional infrastructure.

(3) It uses physics and vehicle dynamics, as well as road user psychology and human factors engineering, to reduce the influence of factors that contribute to crashes.

Ⅴ. Expanding

Benefits of Intelligent Transport Systems.

Intelligent Transport Systems (ITS) have the potential to provide three key benefits for Australian road users of businesses and society.

Safety—Road crashes cause suffering and loss of life as well as costing the nation in the order of US $ 27 billion a year. Many collisions occur due to the stop-start nature of traffic in congested areas. ITS technologies can be used to smooth traffic flows, reduce congestion and hence reduce certain types of accidents. In the future, Cooperative-ITS, which involves communications between vehicles and road-side infrastructure, could be used to improve safety by providing warnings on heavy braking or potential collisions at intersections. Information provided through ITS can also be used to direct traffic away from accidents and alert emergency services as soon as an incident occurs.

Productivity—Congestion lowers productivity, causes flow-on delays in supply chains and increases the cost of business. ITS can increase productivity by finding innovative ways to increase the capacity of our current infrastructure.

Environmental performance—ITS that reduce congestion and stop-start driving can also reduce fuel consumption and greenhouse gas emissions compared with normal driving conditions.

Ⅵ. Reading Material

Urban Transportation Planning

A decision-oriented approach to urban transportation planning should thus focus on the information needs of decision makers and should recognize the often limited capability of individual not familiar with technical analysis to interpret the information that is produced. The underlying assumption of such an approach to planning is that the relevant decision makers can indeed be identified. In the context of urban transportation planning, decision makers are those individuals faced with the problem of allocating resources among competing needs to achieve certain ends. Decision makers can thus include elected officials who set general policies for resource allocation and appropriate funds for the implementation of specific actions, transportation agency managers responsible for operating and maintaining components of the transportation system, private-sector managers who must determine the most efficient routing of urban commodity shipments, and corporate officials concerned with employee transportation.

The approach to urban transportation planning discussed above is based on the information needs of decisions makers. It is important that planning provides not only the information desired by decision makers, but also information needed to provide a more complete understanding of the problem and of the implications of different solutions.

Given these considerations, urban transportation planning can be defined in the following manner. Urban transportation planning is the process of following:

(1) Understanding the types of decisions that need to be made.

(2) Assessing opportunities and limitations of the future.

(3) Identifying the short-and long-term consequences of alternative choices designed

to take advantage of these opportunities or respond to these limitations.

(4) Relating alternative decisions to the goals and objectives established for an urban area, agency, or firm.

(5) Presenting this information to decision makers in a readily understandable and useful form.

Several concepts in this definition merit special attention. First, transportation planning is considered as a process. Such a process includes careful consideration of problem definition, incorporation of alternative viewpoints of analysis and evaluation, development of goals and objective statements, and completion of the technical analysis needed to determine impacts of alternative decisions. It should be emphasized that technical analysis, considered by many to be synonymous with planning, is just one component of the planning process.

Second, transportation planning should assess opportunities as well as limitations of the future. The traditional approach to planning focuses most attention on identifying where "problem" will occur in the transportation system. Clearly, such a focus is an important element of planning. However, there might be opportunities in the existing operation of the transportation system for significant improvements to be made. For example, even though bus routes in an urban transit network might be operation at acceptable performance levels, reorganizing the structure of the service might result in more efficient operations while maintaining, or even improving, service. Planning should thus include a proactive approach to issue definition.

Third, transportation planning should include a long-range and short-range perspective. The long-range planning component is a continuing activity that represents a statement of need and policy direction, thereby providing a context for periodic transportation decisions to be made in the near term. The long-range component of a transportation plan, to be relevant to decision makers, must be both flexible and responsive in scale (level of detail) and scope (alternatives and impacts considered) to the kinds of decisions likely to be made. The short-range component takes into account the more immediate needs of transportation system performance. The relationship between these two components is also a critical concern in the transportation planning process. That is, the extent to which short-term decisions might change the image of future system design and performance (and force or forgo future options), and how anticipated changes in an urban area might influence the effectiveness of shorter-term decisions, are important relationships that need to be addressed in transportation planning.

Fourth, the evaluation of alternative choices is directly related to the goals and objectives established for the planning process. As in Boulding's valuation scheme, goals and objectives form the basis of the measures of effectiveness used in evaluation. Because goals and objectives are intricately tied to values, careful consideration should be given to whose value are represented in a statement of goals and objectives, and efforts should be made to provide community input into the development of such a statement.

Finally, by far the most important decision makers for urban transportation planning are the elected and appointed government officials who must provide and maintain a transportation system that meets the mobility and accessibility needs of their constituents. At the same time, the officials must consider the equity implications of pursuing one program over another. The types of transportation decisions facing these officials include expanding or modifying the existing transportation infrastructure and changing the internal management structure of transportation agencies.

In summary, a decision-oriented approach to urban transportation planning focuses as much attention on the process of planning as it does on the techniques. In some ways, this approach requires the planner to reverse the traditional sequence of planning (i.e., proceeding from problem definition to a final decision) by first understanding the requirements of the final decision and then identifying the information and analysis needed to produce it. The information produced by the planning process can in this way be related to the needs of the decision-making process.

参 考 译 文

第26课 交 通 工 程

交通工程是土木工程的一个分支，其利用工程技术实现人员和货物在道路上安全而高效地流动。它主要关注安全高效的交通流量的研究，如道路几何形状、人行道和人行横道、自行车基础设施、交通标志、路面标记和交通灯。除了提供基础设施外，交通工程还执行运输系统的功能部分。

交通工程还与以下其他学科密切相关：
（1）运输工程；
（2）路面工程；
（3）自行车交通工程；
（4）公路工程；
（5）交通规划；
（6）城市规划；
（7）人力因素工程。

典型的交通工程项目涉及设计交通控制设备的安装与调试，包括交通信号、标志和路面标记。然而，交通工程师也要通过调查高事故率的位置来考虑交通安全并制定减少交通事故的对策。图26-1显示的是有多个车道、自行车道和人行横道的复杂的交叉路口，这些交叉路口是交通工程项目中常见的例子。交通流量管理可以是短期的（准备建设交通管制计划，包括行人和车辆交通的绕行计划）也可以是长期的（评估拟建商业开发对交通模式的影响）。交通问题越来越多地通过开发用于智能交通系统的体系来解决，通常要结合诸如计算机工程和电气工程等其他工程学科来进行。

图 26-1 复杂的交叉路口

1. 交通系统

传统上,道路的改进主要包括建造更多的额外设施。然而,动态元件现在已经被引入到道路交通管理中。动态元件早已用于铁路运输中。这些元件包括用于测量交通流量的传感器和用于管理交通的自动并互连的引导系统(例如,根据一天中的时间节点开放不同方向通行车道的交通标志)。此外,交通流量和速度传感器用于检测问题及提醒操作者,这样就可以确定拥堵的原因,并且可以采取措施将拥堵造成的延迟最小化。这些系统统称为智能交通系统。

2. 车道流量方程

车道流量(Q,每小时车辆数)、最大速度(v,每小时千米数)和密度(K,每千米车辆数)之间的关系是:

$$Q = Kv$$

观察限制进出的设施表明,如果达到最大流量,速度不下降,那么密度就会增加。然而,在临界阈值以上,密度增加会降低速度。此外,进一步超过阈值,密度增加也会降低流量。

因此,通过使用装置来限制高速公路进口处车辆的速率来管控交通密度,可以在高峰期保持瓶颈处较高的速度和车道流量。如图 26-2 所示,匝道调节信号灯,也就是控制车辆进入主道时速率的入口坡道上的信号灯,能够提供此功能(以增加等候在匝道处车辆的等待时间为代价)。

3. 公路安全

公路安全工程是交通工程的一个分支,目的是降低交通事故的频率和严重程度。它应用物理和车辆动力学、道路使用者的心理学及人力因素工程来减少导致事故的影响因素。

典型的交通安全调查遵循以下步骤:

(1)确定调查地点及其优先级。通过查找具有高于平均事故率的地点来选择位置,并解决民众的投诉。

(2)收集数据。这包括获得交警的事故报告,观察道路使用者的行为,以及收集关于交通标志、路面标记、交通信号灯和道路几何形状的信息。

(3)分析数据。寻找可能导致问题的碰撞模式或道路状况。

图 26-2　匝道调节信号灯限制进入高速公路的车辆的速率

（4）确定可能的对策，以减少事故的严重性或频率。

① 评估替代方案的成本/效益比。

② 考虑一个建议的改进措施是否会解决问题或导致"事故迁移"。例如，禁止在一个交叉路口处左转可以消除在该位置处的左转交通事故，结果是使堵车变得更严重。

③ 提出的改进措施的任何缺点可能比您要解决的问题更糟糕？

（5）实施改进。

（6）评估结果。结果通常会在改进措施实施一段时间后得出。事故的严重性和频率已经降到可接受的水平了吗？如果没有，请返回执行步骤(2)。

参 考 文 献

[1] G. D. Taylor. Materials in Construction—An introduction [M]. 3rd ed. London：Routledge，2000.
[2] 赵明瑜．土木建筑系列英语（第四级 工业与民用建筑）[M]．北京：中国建筑工业出版社，1992.
[3] 胡琳，等．Engineering Drawing（工程制图）（英汉双语对照）[M]．北京：机械工业出版社，2010.
[4] John S. Scott. Civil Engineering [M]. London：Longman Group Ltd.，1977.
[5] 段兵廷．土木工程专业英语 [M]．2版．武汉：武汉理工大学出版社，2003.
[6] A C Twort，J Gordon Rees. Civil Engineering Supervision and Management [M]. 3rd ed. London：Arnold，1984.
[7] 王竹芳．工程管理专业英语 [M]．北京：北京大学出版社，2009.
[8] 苏小卒．土木工程专业英语（上册）[M]．2版．上海：同济大学出版社，2011.
[9] 苏小卒．土木工程专业英语（下册）[M]．上海：同济大学出版社，2003.
[10] Mike Riley，Alison Cotgrave. Construction Technology 1—House Construction [M]. London：Palgrave Macmillan，2013.
[11] Francis D. K. Ching. European Building Construction Illustrated [M]. New Jersey：John Willy & Sons，Inc.，2014.
[12] 董祥．土木工程专业英语 [M]．南京：东南大学出版社，2011.
[13] 陈平，王凤池．土木工程基础英语教程 [M]．北京：北京大学出版社，2016.
[14] 史巍，王凯英．土木工程专业英语 [M]．武汉：武汉大学出版社，2014.
[15] 田文玉．土木工程专业英语 [M]．重庆：重庆大学出版社，2010.
[16] 霍俊芳，姜丽云．土木工程专业英语 [M]．北京：北京大学出版社，2013.
[17] 璩继立，宿晓萍．土木工程专业英语 [M]．北京：中国电力出版社，2014.
[18] 刘存中．土木工程专业英语 [M]．北京：机械工业出版社，2008.
[19] Jack Stroud Foster，Roger Greeno. Structure and Fabric—Part 1 [M]. 7th ed. New Jersey：Prentice Hall，200d.
[20] 李著璟．初等钢筋混凝土 [M]．2版．北京：清华大学出版社，2005.

北京大学出版社土木建筑系列教材(已出版)

序号	书名	主编	定价	序号	书名	主编	定价
1	工程项目管理	董良峰 张瑞敏	43.00	50	工程财务管理	张学英	38.00
2	建筑设备(第2版)	刘源全 张国军	46.00	51	土木工程施工	石海均 马哲	40.00
3	土木工程测量(第2版)	陈久强 刘文生	40.00	52	土木工程制图(第2版)	张会平	45.00
4	土木工程材料(第2版)	柯国军	45.00	53	土木工程制图习题集(第2版)	张会平	28.00
5	土木工程计算机绘图	袁果 张渝生	28.00	54	土木工程材料(第2版)	王春阳	50.00
6	工程地质(第2版)	何培玲 张婷	26.00	55	结构抗震设计(第2版)	祝英杰	37.00
7	建设工程监理概论(第3版)	巩天真 张泽平	40.00	56	土木工程专业英语	霍俊芳 姜丽云	35.00
8	工程经济学(第2版)	冯为民 付晓灵	42.00	57	混凝土结构设计原理(第2版)	邵永健	52.00
9	工程项目管理(第2版)	仲景冰 王红兵	45.00	58	土木工程计量与计价	王翠琴 李春燕	35.00
10	工程造价管理	车春鹂 杜春艳	24.00	59	房地产开发与管理	刘薇	38.00
11	工程招标投标管理(第2版)	刘昌明	30.00	60	土力学	高向阳	32.00
12	工程合同管理	方俊 胡向真	23.00	61	建筑表现技法	冯柯	42.00
13	建筑工程施工组织与管理(第2版)	余群舟 宋会莲	31.00	62	工程招投标与合同管理(第2版)	吴芳 冯宁	43.00
14	建设法规(第2版)	肖铭 潘安平	32.00	63	工程施工组织	周国恩	28.00
15	建设项目评估	王华	35.00	64	建筑力学	邹建奇	34.00
16	工程量清单的编制与投标报价	刘富勤 陈德方	25.00	65	土力学学习指导与考题精解	高向阳	26.00
17	土木工程概预算与投标报价(第2版)	刘薇 叶良	37.00	66	建筑概论	钱坤	28.00
18	室内装饰工程预算	陈祖建	30.00	67	岩石力学	高玮	35.00
19	力学与结构	徐吉恩 唐小弟	42.00	68	交通工程学	李杰 王富	39.00
20	理论力学(第2版)	张俊彦 赵荣国	40.00	69	房地产策划	王直民	42.00
21	材料力学	金康宁 谢群丹	27.00	70	中国传统建筑构造	李合群	35.00
22	结构力学简明教程	张系斌	20.00	71	房地产开发	石海均 王宏	34.00
23	流体力学(第2版)	章宝华	25.00	72	室内设计原理	冯柯	28.00
24	弹性力学	薛强	22.00	73	建筑结构优化及应用	朱杰江	30.00
25	工程力学(第2版)	罗迎社 喻小明	39.00	74	高层与大跨建筑结构施工	王绍君	45.00
26	土力学(第2版)	肖仁成 俞晓	25.00	75	工程造价管理	周国恩	42.00
27	基础工程	王协群 章宝华	32.00	76	土建工程制图(第2版)	张黎骅	38.00
28	有限单元法(第2版)	丁科 殷水平	30.00	77	土建工程制图习题集(第2版)	张黎骅	34.00
29	土木工程施工	邓寿昌 李晓目	42.00	78	材料力学	章宝华	36.00
30	房屋建筑学(第3版)	聂洪达	56.00	79	土力学教程(第2版)	孟祥波	34.00
31	混凝土结构设计原理	许成祥 何培玲	28.00	80	土力学	曹卫平	34.00
32	混凝土结构设计	彭刚 蔡江勇	28.00	81	土木工程项目管理	郑文新	41.00
33	钢结构设计原理	石建军 姜袁	32.00	82	工程力学	王明斌 庞永平	37.00
34	结构抗震设计	马成松 苏原	25.00	83	建筑工程造价	郑文新	39.00
35	高层建筑施工	张厚先 陈德方	32.00	84	土力学(中英双语)	郎煜华	38.00
36	高层建筑结构设计	张仲先 王海波	23.00	85	土木建筑CAD实用教程	王文达	30.00
37	工程事故分析与工程安全(第2版)	谢征勋 罗章	38.00	86	工程管理概论	郑文新 李献涛	26.00
38	砌体结构(第2版)	何培玲 尹维新	26.00	87	景观设计	陈玲玲	49.00
39	荷载与结构设计方法(第2版)	许成祥 何培玲	30.00	88	色彩景观基础教程	阮正仪	42.00
40	工程结构检测	周详 刘益虹	20.00	89	工程力学	杨云芳	42.00
41	土木工程课程设计指南	许明 孟苗超	25.00	90	工程设计软件应用	孙香红	39.00
42	桥梁工程(第2版)	周先雁 王解军	37.00	91	城市轨道交通工程建设风险与保险	吴宏建 刘宽亮	75.00
43	房屋建筑学(上:民用建筑)(第2版)	钱坤 王若竹 吴歌	40.00	92	混凝土结构设计原理	熊丹安	32.00
44	房屋建筑学(下:工业建筑)(第2版)	钱坤 吴歌	36.00	93	城市详细规划原理与设计方法	姜云	36.00
45	工程管理专业英语	王竹芳	24.00	94	工程经济学	都沁军	42.00
46	建筑结构CAD教程	崔钦淑	36.00	95	结构力学	边亚东	42.00
47	建设工程招投标与合同管理实务(第2版)	崔东红	49.00	96	房地产估价	沈良峰	45.00
48	工程地质(第2版)	倪宏革 周建波	30.00	97	土木工程结构试验	叶成杰	39.00
49	工程经济学	张厚钧	36.00	98	土木工程概论	邓友生	34.00

序号	书名	主编	定价	序号	书名	主编	定价
99	工程项目管理	邓铁军 杨亚频	48.00	138	建筑学导论	裘鞠 常悦	32.00
100	误差理论与测量平差基础	胡圣武 肖本林	37.00	139	工程项目管理	王华	42.00
101	房地产估价理论与实务	李龙	36.00	140	园林工程计量与计价	温日琨 舒美英	45.00
102	混凝土结构设计	熊丹安	37.00	141	城市与区域规划实用模型	郭志恭	45.00
103	钢结构设计原理	胡习兵	30.00	142	特殊土地基处理	刘起霞	50.00
104	钢结构设计	胡习兵 张再华	42.00	143	建筑节能概论	余晓平	34.00
105	土木工程材料	赵志曼	39.00	144	中国文物建筑保护及修复工程学	郭志恭	45.00
106	工程项目投资控制	曲娜 陈顺良	32.00	145	建筑电气	李云	45.00
107	建设项目评估	黄明知 尚华艳	38.00	146	建筑美学	邓友生	36.00
108	结构力学实用教程	常伏德	47.00	147	空调工程	战乃岩 王建辉	45.00
109	道路勘测设计	刘文生	43.00	148	建筑构造	宿晓萍 隋艳娥	36.00
110	大跨桥梁	王解军 周先雁	30.00	149	城市与区域认知实习教程	邹君	30.00
111	工程爆破	段宝福	42.00	150	幼儿园建筑设计	龚兆先	37.00
112	地基处理	刘起霞	45.00	151	房屋建筑学	董海荣	47.00
113	水分析化学	宋吉娜	42.00	152	园林与环境景观设计	董智 曾伟	46.00
114	基础工程	曹云	43.00	153	中外建筑史	吴薇	36.00
115	建筑结构抗震分析与设计	裴星洙	35.00	154	建筑构造原理与设计(下册)	梁晓慧 陈玲玲	38.00
116	建筑工程安全管理与技术	高向阳	40.00	155	建筑结构	苏明会 赵亮	50.00
117	土木工程施工与管理	李华锋 徐芸	65.00	156	工程经济与项目管理	都沁军	45.00
118	土木工程试验	王吉民	34.00	157	土力学试验	孟云梅	32.00
119	土质学与土力学	刘红军	36.00	158	土力学	杨雪强	40.00
120	建筑工程施工组织与概预算	钟吉湘	52.00	159	建筑美术教程	陈希平	45.00
121	房地产测量	魏德宏	28.00	160	市政工程计量与计价	赵志曼 张建平	38.00
122	土力学	贾彩虹	38.00	161	建设工程合同管理	余群舟	36.00
123	交通工程基础	王富	24.00	162	土木工程基础英语教程	陈平 王凤池	32.00
124	房屋建筑学	宿晓萍 隋艳娥	43.00	163	土木工程专业毕业设计指导	高向阳	40.00
125	建筑工程计量与计价	张叶田	50.00	164	土木工程CAD	王玉岚	42.00
126	工程力学	杨民献	50.00	165	外国建筑简史	吴薇	38.00
127	建筑工程管理专业英语	杨云会	36.00	166	工程量清单的编制与投标报价(第2版)	刘富勤 陈友华 宋会莲	34.00
128	土木工程地质	陈文昭	32.00	167	土木工程施工	陈泽世 凌平平	58.00
129	暖通空调节能运行	余晓平	30.00	168	特种结构	孙克	30.00
130	土工试验原理与操作	高向阳	25.00	169	结构力学	何春保	45.00
131	理论力学	欧阳辉	48.00	170	建筑抗震与高层结构设计	周锡武 朴相顺	36.00
132	土木工程材料习题与学习指导	鄢朝勇	35.00	171	建设法规	刘红霞 柳立生	36.00
133	建筑构造原理与设计(上册)	陈玲玲	34.00	172	道路勘测与设计	凌平平 余婵娟	42.00
134	城市生态与城市环境保护	梁彦兰 阎利	36.00	173	工程结构	金恩平	49.00
135	房地产法规	潘安平		174	建筑公共安全技术与设计	陈继斌	45.00
136	水泵与水泵站	张伟 周书葵	35.00	175	地下工程施工	江学良 杨慧	54.00
137	建筑工程施工	叶良	55.00	176	土木工程专业英语	宿晓萍 赵庆明	40.00

如您需要更多教学资源如电子课件、电子样章、习题答案等，请登录北京大学出版社第六事业部官网 www.pup6.cn 搜索下载。

如您需要浏览更多专业教材，请扫下面的二维码，关注北京大学出版社第六事业部官方微信（微信号：pup6book），随时查询专业教材、浏览教材目录、内容简介等信息，并可在线申请纸质样书用于教学。

感谢您使用我们的教材，欢迎您随时与我们联系，我们将及时做好全方位的服务。联系方式：010-62750667，donglu2004@163.com，pup_6@163.com，lihu80@163.com，欢迎来电来信。客户服务 QQ 号：1292552107，欢迎随时咨询。